T0004097

BARRON'S

Physics Practice Plus

400+ ONLINE QUESTIONS

WITH QUICK REVIEW GUIDE

ROBERT JANSEN, M.A., AND GREG YOUNG, M.S.Ed.

About the Authors

Robert Jansen

Robert Jansen has taught Advanced Placement Physics at Aliso Niguel High School in Aliso Viejo, California, since 1998. He holds a bachelor's degree in psychobiology from the University of California, Los Angeles, and a master's degree in education from Pepperdine University. He gravitated toward teaching physics because of the challenging material and a sustained belief that physics does not need to be mysterious and difficult but rather can be comprehensible and achievable. The result has been a large and competitive physics program where each year more than 220 students participate in the AP Physics 1 and AP Physics C courses. The overarching goal is preparedness and confidence for students who will be studying science during their undergraduate years.

Greg Young

Greg Young has been teaching high school science for more than twenty years. He currently teaches Honors Physics, AP Chemistry, and Chemistry at San Clemente High School in San Clemente, California. He holds a bachelor's degree in biochemistry from the University of California, San Diego, and a master's degree in science education from USC. Having always been interested in science and how to make it relevant to others, Greg's interest in teaching lies in being able to create interactive lessons that engage students in their learning and form a relevant context for difficult concepts in physics and chemistry. Science made interesting is science worth learning.

© Copyright 2022 by Kaplan North America, LLC, dba Barron's Educational Series. Content in this book was previously published in a different format under the title *SAT Subject Test in Physics*.

Published by Kaplan North America, LLC, dba Barron's Educational Series
1515 West Cypress Creek Road
Fort Lauderdale, FL 33309
www.barronseduc.com

All rights reserved. No part of this publication may be reproduced or distributed in any form or by any means without the written permission of the copyright owner.

10 9 8 7 6 5 4 3 2 1

ISBN: 978-1-5062-8152-0

Kaplan North America, LLC, dba Barron's Educational Series print books are available at special quantity discounts to use for sales promotions, employee premiums, or educational purposes. For more information or to purchase books, please call the Simon & Schuster special sales department at 866-506-1949.

CONTENTS

How to Use This Book vii

CHAPTER 1: CONVENTIONS AND GRAPHING. 1

Fundamental and Derived Units. 1

Graphing Variables 2

Slope and Area 3

Interpreting Graphs 5

CHAPTER 2: VECTORS 7

Coordinate System 7

Scalars. 8

Vectors 8

Vector Mathematics 11

CHAPTER 3: KINEMATICS IN ONE DIMENSION. 16

Kinematic Quantities 16

Kinematic Equations 19

Kinematic Graphs. 21

CHAPTER 4: KINEMATICS IN TWO DIMENSIONS. 23

Independence of Motion. 23

True Velocity and Displacement 24

Relative Velocity 25

Projectile Motion 27

Projectiles Launched at an Angle 29

CHAPTER 5: DYNAMICS 30

Inertia 31

Force. 31

Common Forces 31

Force Diagrams 35

Newton's Laws of Motion 36

Solving Force Problems 38

CHAPTER 6: CIRCULAR MOTION. 46

Uniform Circular Motion 47

Dynamics in Circular Motion. 50

Angular Velocity 51

CHAPTER 7: ENERGY, WORK, AND POWER 52

Mechanical Energy 52

Work 55

Power 61

Conservation of Energy 62

CHAPTER 8: MOMENTUM AND IMPULSE 64

Momentum 64

Impulse 65

Conservation of Momentum. 67

Energy in Collisions. 67

CHAPTER 9: GRAVITY. 70

Universal Gravity 70

Gravitational Field 71

Circular Orbits. 72

Kepler's Laws 74

CHAPTER 10: ELECTRIC FIELDS 75

Charge. 76

Electric Fields 77

Uniform Electric Fields 77

Electric Fields of Point Charges 80

CHAPTER 11: ELECTRIC POTENTIAL 85

Potential of Uniform Fields 86

Potential of Point Charges. 88

Electric Potential Energy 89

Motion of Charges and Potential 90

Capacitors. 92

CHAPTER 12: CIRCUIT ELEMENTS AND DC CIRCUITS 95

Principal Components of a DC Circuit 95

DC Circuits 97

Heat and Power Dissipation. 102

CHAPTER 13: MAGNETISM 104

Permanent or Fixed Magnets 105

Current-Carrying Wires 106

Solenoids and Electromagnets 108

Force on Moving Charges 108

Force on Current-Carrying Wires 112

Electromagnetic Induction 113

CHAPTER 14: SIMPLE HARMONIC MOTION 117

Terms Related to SHM 117

Oscillations of Springs 118

Oscillations of Pendulums 121

Graphical Representations of SHM 122

CHAPTER 15: WAVES 124

Traveling Waves 124

Mechanical Waves 127

Electromagnetic Waves 127

Doppler Effect 128

Superposition and Standing Waves 131

CHAPTER 16: GEOMETRIC OPTICS 135

Ray Model of Light 136

Reflection 136

Refraction 137

Thin Lenses 139

Spherical Mirrors 144

CHAPTER 17: PHYSICAL OPTICS 148

Diffraction 148

Interference of Light 150

Polarization of Light 152

Color 154

CHAPTER 18: THERMAL PROPERTIES .156

Thermal Systems 156

Thermal Energy 157

Temperature 158

Thermal Expansion 158

Ideal Gases 160

Heat and Heat Transfer 162

Heating and Cooling 164

CHAPTER 19: THERMODYNAMICS. .167

Internal Energy 167

Energy Transfer in Thermodynamics 168

Energy Model Summarized 170

First Law of Thermodynamics. 170

Entropy. 171

Second Law of Thermodynamics 172

CHAPTER 20: ATOMIC AND QUANTUM PHENOMENA 173

Development of the Atomic Theory 173

Energy-Level Transitions 177

Ionization Energy/Work Function 179

Photoelectric Effect 179

CHAPTER 21: NUCLEAR REACTIONS 181

Nucleons 181

Subatomic Particles 182

Isotopes 184

The Strong Force 184

Mass-Energy Equivalence 185

Radioactive Decay 186

Fission and Fusion 188

CHAPTER 22: RELATIVITY. 190

Special Theory of Relativity 190

Time, Length, and Mass 191

INDEX. 193

How to Use This Book

Barron's Physics Practice Plus is designed to offer essential review of key topics and loads of online practice to help you excel in physics.

Online Practice

Access more than 400 questions in online quizzes arranged by topic for customized practice! All questions include answer explanations.

What Will You Learn in the Book?

Key review and topics are covered so you can study the essentials needed to succeed.

Learning objectives are listed at the start of each chapter. This list of key ideas will help guide your learning and study plan and allow you to easily return to topics that you want to review again.

Tips are given throughout the book to offer helpful notes, reminders, and strategies to improve your learning.

CHAPTER 1

Conventions and Graphing

Learning Objectives

In this chapter, you will learn how to:

- Review the fundamental metric units (SI units) and some of the derived metric units (SI units) used in physics
- Determine the dependent and independent variables of a graph
- Explain the importance of slope and area to a graph

Fundamental and Derived Units

The **fundamental metric units** (SI units) in physics cover the basic quantities measured, such as length, mass, and time. The units measure a quantity and are given a unit name and symbol. Table 1.1 lists the fundamental quantities along with the unit names and symbols.

TABLE 1.1 Fundamental Quantities and Units

Quantity (Symbol)	Unit Name	Symbol
Length (l)	Meter	m
Mass (m)	Kilogram	kg
Time (t)	Second	s
Electric current (I)	Ampere	A
Temperature (T)	Kelvin	K
Amount of substance (n)	Mole	mol

Derived units are combinations of one or more of the fundamental units. Table 1.2 lists common derived units used in physics.

TABLE 1.2 Derived Units

Quantity (Symbol)	Unit Name	Unit Symbol	Fundamental Units
Area (A)	Area	m^2	m^2
Volume (V)	Volume	m^3	m^3
Density (ρ)	Density	kg/m^3	kg/m^3
Frequency (f)	Hertz	Hz	$1/s = s^{-1}$
Force (F)	Newton	N	$kg \cdot m/s^2$
Energy (E)	Joule	J	$N \cdot m = kg \cdot m^2/s^2$
Power (P)	Watt	W	$J/s = kg \cdot m^2/s^3$
Pressure (P)	Pascal	Pa	$N/m^2 = kg/m \cdot s^2$
Electric charge (q)	Coulomb	C	$A \cdot s$
Electric potential (V)	Volt	V	$J/C = J/A \cdot s = kg \cdot m^2/A \cdot s^3$

It is important to know which units correctly belong to a specific quantity. An easy way to do this is to write out the principal formula for the quantity and then replace each variable on the right side of the equation with its unit symbol. There may be more than one correct answer including the unit symbol, other derived units, and fundamental units. For example, all of the following are correct ways to express units of energy: J, N • m, and kg • m^2/s^2.

EXAMPLE 1.1

Derived Units
The unit of force is the newton. What are the fundamental units that make up the newton?

WHAT'S THE TRICK?
Write down the foundational formula for force.

$$\vec{F} = m\vec{a}$$

Replace the variable symbols with their matching units. Force is measured in newtons, N. Mass is measured in kilograms, kg. Acceleration is measured in meters per second squared, m/s^2.

$$N = kg \cdot m/s^2$$

Graphing Variables

The graphing techniques of mathematics are used in science to compare dependent and independent variables. In mathematics, you are familiar with the traditional x- and y-coordinate axes. In science, the x-axis represents the independent variable and the y-axis represents the dependent variable. The value of the dependent variable depends upon the independent variable.

Graphs are always titled so that the dependent variable is listed first, and the independent variable is listed second. As an example, a position versus time graph would have position (dependent variable) plotted on the y-axis and time (independent variable) plotted on the x-axis.

Slope and Area

Slope

Slopes are very important in physics. Slope is determined by dividing the rise (y-axis value) by the run (x-axis value). The trick is to look at the units written on the axes of the graph. If you divide these units, you can easily identify the significance of the slope.

EXAMPLE 1.2

Slope of a Graphed Function

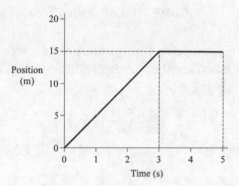

(A) What is the value and significance of the slope in the time interval from 0 to 3 seconds?

WHAT'S THE TRICK?

Determining the slope is simply a matter of dividing the rise (y-axis values) by the run (x-axis values). The significance of the slope is determined by examining the resulting units.

$$\text{slope} = \frac{\text{rise}}{\text{run}} = \frac{15m - 0m}{3s - 0s} = 5 \text{ m/s}$$

The resulting units, meters per second (m/s), are the units of velocity. Therefore, the slope of the position versus time graph is equal to velocity. During the first 3 seconds, the object has a velocity of 5 m/s.

(B) What is the value and significance of the slope in the time interval from 3 to 5 seconds?

WHAT'S THE TRICK?

The slope in the interval between 3 and 5 seconds is zero.

$$\text{slope} = \frac{\text{rise}}{\text{run}} = \frac{15m - 15m}{5s - 3s} = 0 \text{ m/s}$$

During this time interval, the object has a velocity of zero and the y-axis value (position) is not changing. The object's position remains constant at a location 15 m from the origin.

Area

The **area** formed by the boundary between the x-axis and the line of a graph is also very useful. Areas are calculated by multiplying the height (y-axis value) by the base (x-axis value). In problems where the area forms a triangle, the area is found with $\frac{1}{2}$ height × base. In cases where the line of the graph is below the x-axis, the area is negative. See Figure 1.1.

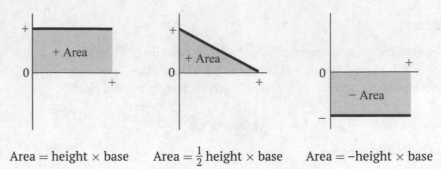

$$Area = height \times base \qquad Area = \frac{1}{2} height \times base \qquad Area = -height \times base$$

FIGURE 1.1 Calculating area

As with slope, you can easily determine the significance of the area. By multiplying the units written on the axes of the graph and then looking at the resulting units, you can quickly determine the significance of the area.

EXAMPLE 1.3

Area of a Graphed Function

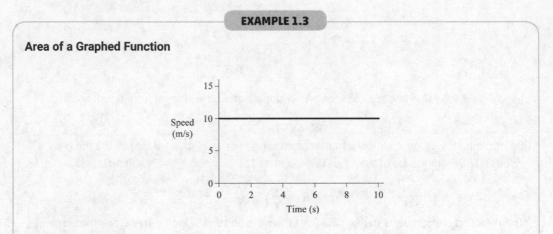

What is the value and significance of the area of the graph during the time interval between 0 and 10 seconds?

WHAT'S THE TRICK?

Determine the area, and examine the resulting units.

$$area = height \times base = (10 \text{ m/s})(10 \text{ s}) = 100 \text{ m}$$

Meters (m) are the units of displacement. The area under a speed versus time graph is therefore the displacement of the object during that time interval. The object graphed above traveled 100 meters in 10 seconds.

Interpreting Graphs

Consider the graph of velocity versus time in Figure 1.2.

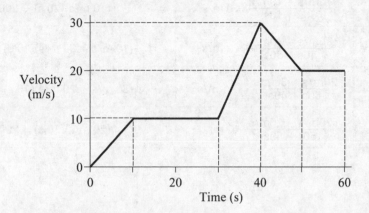

FIGURE 1.2 Velocity versus time graph

The graph tells the story of an object, such as a car, as it moves over a 60-second period of time. At time zero, the object has a velocity of 0 meters per second and is therefore starting from rest. The y-intercept of a speed versus time graph is the initial velocity of the object, v_0.

What the object is doing during the 60 seconds can be determined by analyzing the slope and area during the separate time intervals. Determine the significance of the slope by dividing the rise units (y-axis values) by the run units (x-axis values).

$$\text{slope units} = \frac{\text{rise units}}{\text{run units}} = \frac{\text{m/s}}{\text{s}} = \text{m/s}^2$$

The slope units, meters per second squared (m/s^2), are the units of acceleration. Thus, the slope of speed versus time is acceleration. Determine the significance of the area between the graphed function and the x-axis by multiplying the units of the y-axis by the units of the x-axis.

$$\text{area units} = \text{height units} \times \text{base units} = \frac{\text{m}}{\text{s}} \times \text{s} = \text{m}$$

Meters (m) are the units of displacement. The area of a velocity versus time graph is displacement.

To analyze the motion mathematically, divide the graph into a series of line segments and evaluate each section. The following chart shows the acceleration and displacement for the time intervals corresponding to the graphed line segments.

Time (s)	Slope (Acceleration)	Area (Displacement)
0 to 10	$\dfrac{10\text{m/s} - 0\text{m/s}}{10\text{s} - 0\text{s}} = 1 \text{ m/s}^2$	$\frac{1}{2}(10 \text{ m/s})(10 \text{ s}) = 50 \text{ m}$
10 to 30	$\dfrac{10\text{m/s} - 10\text{m/s}}{30\text{s} - 10\text{s}} = 0 \text{ m/s}^2$	$(10 \text{ m/s})(20 \text{ s}) = 200 \text{ m}$
30 to 40	$\dfrac{30\text{m/s} - 10\text{m/s}}{40\text{s} - 30\text{s}} = 2 \text{ m/s}^2$	$\frac{1}{2}(10 \text{ m/s} + 30 \text{ m/s})(10 \text{ s}) = 200 \text{ m}$
40 to 50	$\dfrac{20\text{m/s} - 30\text{m/s}}{50\text{s} - 40\text{s}} = -1 \text{ m/s}^2$	$\frac{1}{2}(30 \text{ m/s} + 20 \text{ m/s})(10 \text{ s}) = 250 \text{ m}$
50 to 60	$\dfrac{20\text{m/s} - 20\text{m/s}}{60\text{s} - 50\text{s}} = 0 \text{ m/s}^2$	$(20 \text{ m/s})(10 \text{ s}) = 200 \text{ m}$

CHAPTER 2

Vectors

Learning Objectives

In this chapter, you will learn how to:

○ Identify a mathematical coordinate system that will provide a common frame of reference to orient direction in physics problems

○ Understand the differences and similarities among scalar and vector quantities

○ Resolve vectors into components and add vector quantities

Coordinate System

Problems in physics often involve the motion of objects. Position, displacement, velocity, and acceleration are key numerical quantities needed to describe the motion of an object. Position involves a specific location, while velocity and acceleration act in specific directions. Using the mathematical **coordinate system** is ideal to visualize both position and direction. The coordinate system provides a common frame of reference in which the quantities describing motion can be easily and consistently compared with one another.

We can place an axis anywhere, and we can orient the axis in any direction of our choosing. If a problem does not specify a starting location or direction, then position the origin at the object's starting location. In Figure 2.1, a problem involving the motion of a car can be visualized as starting at the origin and moving horizontally along the positive x-axis.

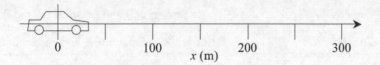

FIGURE 2.1 Horizontal motion

In more complex problems, some quantities cannot be oriented along a common axis. In these problems, direction must be specified in degrees measured counterclockwise (ccw) from the positive x-axis, as shown in Figure 2.2.

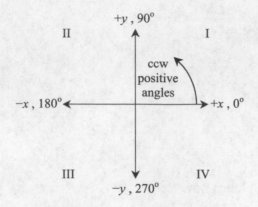

FIGURE 2.2 Coordinate system

A coordinate system is a valuable tool that provides a frame of reference when position and direction are critical factors.

Scalars

A **scalar** is a quantity having only a numerical value. No direction is associated with a scalar. The numerical value describing a scalar is known as its magnitude. Some examples of commonly used scalars are listed in Table 2.1.

TABLE 2.1 Commonly Used Scalars

Quantities Involving	Examples of Common Scalars
Time	Time—t, period—T, and frequency—f
Motion	Distance—d and speed—v
Energy	Kinetic energy—K or KE and potential energy—U or PE
Mass	Mass—m and density—ρ
Gases	Pressure—P, volume—V, and temperature—T
Electricity	Charge—q or Q and potential (voltage)—V
Circuits	Current—I, resistance—R, and capacitance—C

The symbols representing scalars are printed in *italics*. For example, a mass of 2.0 kilograms will be written as $m = 2.0$ kg. Scalars can have magnitudes that are positive, negative, or zero. For example, time = 60 seconds, speed = 0 meters per second, and temperature = $-10°C$.

Vectors

Although scalars possess only magnitude, **vectors** possess both magnitude and a specific direction. Examples of commonly encountered vectors are listed in Table 2.2.

TABLE 2.2 Commonly Used Vectors

Vector Quantity	Vector Symbol	Component Symbol
Displacement	$\vec{s}$ or $\vec{r}$	x or y
Velocity	$\vec{v}$	v_x or v_y
Acceleration	$\vec{a}$	a_x or a_y
Force	$\vec{F}$	F_x or F_y
Momentum	$\vec{p}$	p_x or p_y
Electric field	$\vec{E}$	E_x or E_y
Magnetic field	$\vec{B}$	B_x or B_y

Formal vector variables are usually written in italics with a small arrow drawn over the letter, as shown in the middle column in Table 2.2. You may encounter vector quantities, such as force, in any one of these forms: $\vec{F}$, $\mathbf{F}$, F_x, F_y, and F. The first, $\vec{F}$, is the most accepted and distinctly indicates a vector quantity. It is the format used in this book. The second, $\mathbf{F}$, is an alternate way to indicate a vector quantity. The next two, F_x and F_y, signify vector components that lie along the specific axis indicated by their subscripts. The last, F, appears to be the convention to indicate a scalar quantity. It is typically used when only the magnitude of the vector is needed and the direction is understood.

Distinguishing between vectors and scalars by simply looking at an equation can be confusing. How, then, do you tell scalars and vectors apart? Physics problems may contain clues in the text of the problem to help distinguish vectors from scalars. The mention of a specific direction definitely indicates a vector quantity. However, it is up to you to learn which quantities are vectors and when the use of vector components is necessary. Counting on the use of a specific set of symbol conventions may not be wise.

Vectors do follow certain mathematical conventions that are worth noting. Vector magnitudes can be only positive or zero. However, vectors can have negative direction. Consider the acceleration of gravity, a vector quantity acting in the negative y-direction. The gravity vector includes both magnitude and direction ($\vec{g} = 10$ m/s^2, $-y$). Substituting this exact expression, including the negative y-direction, into an equation is not really workable. Instead the value -10 m/s^2 may be substituted into equations. The negative sign in front of the magnitude indicates the negative y-direction. This can be done only if all the vector quantities used in an equation lie along the same axis and it is understood that the signs on all vector quantities represent direction along that axis. This essentially transforms the vector quantities into scalar quantities, allowing normal mathematical operations. As a result, the variable may be shown as a scalar in italics ($g = -10$ m/s^2) rather than in bold print. When a negative sign is associated with a vector quantity, it technically specifies the vector's direction and assists with proper vector addition.

Vectors are represented graphically as arrows. For displacement vectors, the tail of the arrow is the initial position of the object, x_i, and the tip of the arrow is the final position of the object, x_f. The length of the arrow represents the vector's magnitude, and its orientation on the coordinate axis indicates direction. This may give some insight into the reason that some vector quantities are displayed in italics.

Figure 2.3 shows a car moving 200 meters and its associated vector.

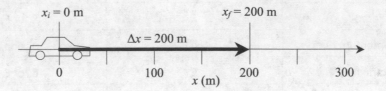

FIGURE 2.3 Horizontal displacement

The magnitude of the displacement vector, Δx, is the absolute value of the difference between the final position, x_f, and the initial position, x_i. Direction can be seen in the diagram.

$$\Delta x = x_f - x_i = 200 - 0 = 200 \text{ m, to the right } (+x)$$

For other vectors, such as velocity and force, the quantity described by the vector occurs at the tail of the arrow. The tail of the arrow shows the actual location of the object being acted upon by the vector quantity. The tip of the arrow points in the direction the vector is acting. The length of the arrow represents the magnitude of the vector quantity. The magnitude and direction described by these types of vectors may be instantaneous values capable of changing as the object moves. In addition, the object may not reach the location specified by the tip of the arrow.

These types of vectors are readily seen in projectile motion. In Figure 2.4, a projectile is launched with a speed of 50 meters per second at an angle of 37° above the horizontal.

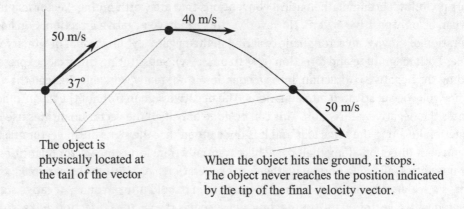

FIGURE 2.4 Projectile motion

Although only three key velocity vectors are shown in the diagram, they clearly demonstrate how the magnitude and direction of velocity change throughout the flight. During the motion depicted in the diagram, no two instantaneous velocity vectors are completely alike.

Knowing how to recognize vectors quantities like displacement, velocity, acceleration, and force will improve your problem-solving skills. The importance of vector direction cannot be overstated. Including the correct sign representing a vector's direction is often the key to arriving at the correct solution. The next sections will demonstrate the importance of vector direction as we review basic vector mathematics.

Vector Mathematics

Components

Vectors aligned to the x- and y-axes are mathematically advantageous. However, some problems involve diagonal vector quantities. Diagonal vectors act simultaneously in both the x- and y-directions, and they are difficult to manipulate mathematically. Fortunately, diagonal vectors can be resolved into x- and y-component vectors. The x- and y-component vectors form the adjacent and opposite sides of a right triangle where the diagonal vector is its hypotenuse. Aligning the component vectors along the x- and y-axes simplifies vector addition.

The magnitudes of component vectors are determined using right-triangle trigonometry. In Figure 2.5, vector **A** is a diagonal vector. It has a magnitude of A and a direction of θ.

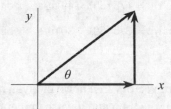

FIGURE 2.5 Magnitudes of vectors

Vector A_x is the x-component of $\vec{A}$ and is adjacent to angle θ. Vector A_y is the y-component of $\vec{A}$ and is opposite angle θ. Normally, the magnitude of the components of vector $\vec{A}$ would be determined using the following right-triangle trigonometry.

$$A_x = A \cos \theta, +x\text{-direction}$$

$$A_y = A \sin \theta, +y\text{-direction}$$

EXAMPLE 2.1

Determining Component Vectors

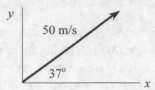

A projectile is launched with an initial velocity of 50 meters per second at an angle of 37° above the horizontal. Determine the *x*- and *y*-component vectors of the velocity.

WHAT'S THE TRICK?

Draw the component vectors and identify the adjacent and opposite sides.

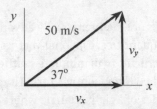

Determining the magnitudes of each component requires multiplying the hypotenuse by the correct fraction. The direction of each component can be determined by looking at the diagram.

$$v_x = 4/5 \text{ hyp} = 4/5 \ (50) = 40 \text{ m/s}, +x\text{-direction}$$

$$v_x = 3/5 \text{ hyp} = 3/5 \ (50) = 30 \text{ m/s}, +y\text{-direction}$$

In some problems, the component vectors are known or given and you must determine the vector they describe. Pythagorean theorem and inverse tangent are used to calculate the magnitude and direction of the diagonal vector described by the component vectors.

$$|\vec{A}| = \sqrt{A_x^2 + A_y^2} \text{ and } \tan \theta = \frac{A_y}{A_x}$$

Adding Vectors

One important aspect of working with vectors is the ability to add two or more vectors together. Only vectors with the same units for magnitude can be added to each other. The result of adding vectors together is known as the vector sum, or resultant.

You can use two visual methods to add vectors. The first is the tip-to-tail method, and the second is the parallelogram method. In some problems, the resultant is known or given and you must determine the magnitude and direction of one of the vectors contributing to the vector sum. The sections below detail examples of each of these scenarios.

Tip-to-Tail Method

Adding vectors **tip to tail** is advantageous when a vector diagram is not given. Begin by sketching a coordinate axis. Vectors can be added in any order. However, drawing *x*-direction vectors first, followed by *y*-direction vectors, is best. Choose the first vector and draw it starting from the origin and pointing in the correct direction. Start drawing the tail of the next vector at the tip of the previous vector. Keep the orientation of the second vector the same as it was given in the problem. Continue this process, adding any remaining vectors to the tip of each subsequent vector. Finally, draw the resultant vector from the origin (tail of the first vector) pointing to the tip of the last vector. You will encounter three common cases of vector addition.

- Vectors pointing in the same direction
- Vectors pointing in opposite directions
- Vectors that are 90° apart

EXAMPLE 2.2

Adding Vectors Pointing in the Same Direction
A person walks 40 meters in the positive x-direction, pauses, and then walks an additional 30 meters in the positive x-direction. Determine the magnitude and direction of the person's displacement.

WHAT'S THE TRICK?
When vectors point in the same direction, simply add them together. Sketch or visualize the vectors tip to tail. The resultant is equal to the total length of both vectors added together.

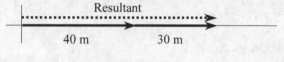

Resultant = 40 m + 30 m = 70 m

EXAMPLE 2.3

Adding Vectors Pointing in Opposite Directions
A person walks 40 meters in the positive x-direction, pauses, and then walks an additional 30 meters in the negative x-direction. Determine the magnitude and direction of the person's displacement.

WHAT'S THE TRICK?
When a vector points in the opposite (negative) direction, you can insert a minus sign in front of the magnitude. Technically, vectors cannot have negative magnitudes. The minus sign actually indicates the vector's direction, and it represents a vector turned around 180°. Again, sketching or visualizing the vectors tip to tail will help you arrive at the correct resultant. The resultant is drawn from the origin to the tip of the last vector added.

Resultant = 40 m + (−30 m) = 10 m

EXAMPLE 2.4

Adding Vectors That Are 90° Apart

An object moves 100 meters in the positive x-direction and then moves 100 meters in the positive y-direction. Determine the magnitude and direction of the object's displacement.

WHAT'S THE TRICK?

Start at the origin and draw the x-direction vector first. Then, add the tail of the y-direction vector to the tip of the first vector. Finally, draw the resultant from the origin pointing toward the tip of the final vector added.

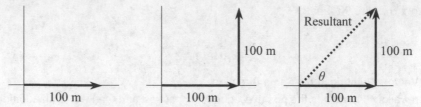

The components and resultant form a 45°-45°-90° triangle. The magnitude of the hypotenuse can be obtained by multiplying a side by the square root of two.

$$\text{hypotenuse} = (\sqrt{2})(\text{side}) = (\sqrt{2})(100), \theta = 45°$$

Without a calculator, $100\sqrt{2}$ is the mathematically simplified answer.

Parallelogram Method

Sometimes, a vector diagram may be provided that shows the vectors in a tail-to-tail configuration. You can add these vectors by constructing a **parallelogram**, as shown in the example below.

EXAMPLE 2.5

Adding Vectors Using the Parallelogram Method

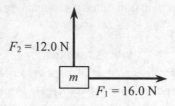

A mass, m, is acted upon by two force vectors, $\vec{F}_1 = 16.0$ N in the +x-direction and $\vec{F}_2 = 12.0$ N in the +y-direction, as shown in the diagram above. Determine the magnitude and direction of the resultant force acting on mass m.

WHAT'S THE TRICK?

Construct a parallelogram. The diagram below on the left shows a dashed line drawn from the tip of $\vec{F}_1$ parallel to $\vec{F}_2$ and a second dashed line drawn from the tip of $\vec{F}_2$ parallel to $\vec{F}_1$. In the diagram below on the right, the resultant is drawn with its tail starting at the origin and the tip extending to the intersection of the dashed lines. The resultant is the sum of the force vectors ($\Sigma\vec{F}$).

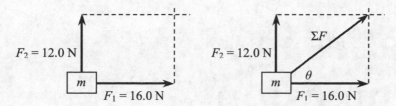

The dashed lines have the same length as the given vectors. Adding the resultant to the diagram creates two right triangles. Look carefully at the ratio of the sides. Two 3-4-5 triangles have been formed.

$$\Sigma\vec{F} = 20.0 \text{ N at } 37°$$

Finding a Missing Vector

In some problems, the resultant is known and the problem requires you to find the magnitude and direction of a missing vector. This frequently occurs when clues in the problem lead you to the conclusion that the resultant vector has a magnitude equal to zero. In order for two vectors to add up to zero, the vectors must have equal magnitudes and point in opposite directions.

EXAMPLE 2.6

Deducing the Existence of a Missing Vector

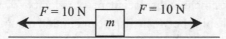

A mass, m, is initially at rest on a horizontal surface. A 10-newton force acting in the positive x-direction is applied to mass m. The mass remains at rest. Why?

WHAT'S THE TRICK?

A force is either a push or a pull. When an object remains stationary, all the pushing forces acting on the object must cancel out each other. Therefore, the sum of all the force vectors is zero. You must conclude that a second force is acting on the mass to cancel the force given in the problem. The only force capable of canceling the given force is a 10-newton force acting in the opposite direction.

CHAPTER 3

Kinematics in One Dimension

Learning Objectives

In this chapter, you will learn how to:

○ Discuss and compare the kinematic quantities

○ Identify signs used for kinematic variables

○ Apply the correct kinematic equation to solve problems

○ Interpret graphical representations of kinematic equations

Table 3.1 lists the variables and their units that are used in a study of kinematics.

TABLE 3.1 Variables and Units Used in Kinematics

Variables Used in Kinematics	Units
$\vec{x}$ = Displacement (distance) (Also: $\Delta x, y, \Delta y, h, \Delta h$, or d)	m (meters)
$\vec{v}_i$ = Initial velocity (speed)	m/s (meters per second)
$\vec{v}_f$ = Final velocity (speed)	m/s (meters per second)
$\vec{a}$ = Acceleration	m/s^2 (meters per second squared)
t = Elapsed time	s (seconds)

Kinematic Quantities

Kinematics involves the mathematical relationship among key quantities describing the motion of an object. These quantities include displacement—$\vec{x}$, velocity—$\vec{v}$, and acceleration—$\vec{a}$. You should also note the relationships between displacement and distance and between velocity and speed.

Displacement and Distance

Displacement, $\vec{x}$, is a vector extending from the initial position of an object to its final position. The variable x is typically used for horizontal motion, while y and h (height) are used for vertical motion.

Displacement differs slightly from distance, d. **Distance** is a scalar quantity representing the actual path followed by the object. When an object travels in a straight line and does not reverse its direction, then distance and the magnitude of displacement are interchangeable.

Velocity and Speed

Velocity and speed are kinematic quantities measuring the rate of change in displacement and distance. A rate is a mathematical relationship showing how one variable changes compared with another. When the word *rate* appears in a problem, simply divide the quantity mentioned by time. **Velocity**, $\vec{v}$, is a vector describing the rate of displacement, $\Delta\vec{x}$. The equation for velocity

$$\vec{v}_{avg} = \frac{\Delta\vec{x}}{t}$$

solves for the average velocity during a time interval, t. Additional information is needed to determine if velocity is constant or is changing during the time interval.

If velocity is changing, then it has different values at different moments in time. However, instantaneous velocity is the velocity at a specific time, t. If you report that you are driving north at 65 mph, you have given an instantaneous velocity. This is a snapshot, freezing the problem at a specific instant. In kinematics, you will encounter two specific instantaneous velocities. Initial velocity, v_i, is at the start of a problem. Final velocity, v_f, is at the end of a problem.

Velocity is a vector quantity, so it includes a specific magnitude and a direction. When the magnitude and direction of velocity are both constant, we say that the object is moving at constant velocity. However, when direction is changing, the term *speed* may be used. **Speed** is a scalar quantity that calculates the rate of distance, as opposed to displacement. If an object travels in a straight line, then the terms *speed* and *velocity* are interchangeable.

Acceleration

Acceleration is the rate of change in velocity.

$$\vec{a} = \frac{\Delta\vec{v}}{t}$$

When acceleration is uniform, its magnitude remains constant. The magnitude of acceleration indicates how quickly velocity is changing. In other words, acceleration is the rate of a rate, which is why students new to physics often have difficulty comprehending it. The effect of acceleration on velocity depends on the orientation of the vector quantities relative to one another, as shown in Table 3.2.

TABLE 3.2 The Effect of Acceleration on Velocity

Same Direction	Opposite Direction	Perpendicular Direction
Acceleration in the same direction as initial velocity causes an increase in speed.	Acceleration in the opposite direction as initial velocity causes a decrease in speed.	Acceleration perpendicular to initial velocity causes a change in direction.

The possibility of changing direction is often overlooked. A car moving around a circular track at a constant speed is said to have uniform acceleration as opposed to constant acceleration. The phrase "uniform acceleration" indicates that the magnitude of acceleration will remain the same while the direction may be changing. The phrase "constant acceleration," however, indicates that both magnitude and direction are the same.

Acceleration of Gravity

All objects on Earth are subject to the acceleration of gravity. This acceleration has a known value at Earth's surface. It is so prevalent in physics problems that it receives its own variable, g. The acceleration of gravity acts downward and has a value of 9.8 m/s^2.

Sign Conventions

The kinematic quantities—displacement, velocity, and acceleration—are all vectors. The magnitude of a vector is always positive. However, vectors can point in either a positive or a negative direction. When vectors point in a negative direction, a negative sign is added to the magnitude of the vector for calculation purposes.

The coordinate-axis system is the best tool to use when determining the correct sign on vector quantities. Picture the object at the origin of the coordinate axes at the start of the problem ($x_i = 0$ and $y_i = 0$). The default positive directions are right and upward. Any vector quantities pointing to the left or downward will include a negative sign in calculations. Table 3.3 summarizes the sign conventions for kinematic variables.

TABLE 3.3 Signs Used for Kinematic Variables

Variable and Sign	Horizontal Motion	Vertical Motion
$+x$	Object finishes right of starting point	Object finishes above starting point
$-x$	Object finishes left of starting point	Object finishes below starting point
$+v$	Moving right	Moving upward
$-v$	Moving left	Moving downward
$+a$	Positive acceleration increases the speed of objects that have a positive velocity. However, when an object has a negative velocity, a positive acceleration acts to decrease speed.	
$-a$	Negative acceleration decreases the speed of objects that have a positive velocity. However, when an object has a negative velocity, a negative acceleration acts to increase speed.	In free-fall problems, acceleration is equal to g and always acts downward.

As seen in the table above, the sign on acceleration can have opposing effects depending on the sign of velocity. If acceleration and velocity vectors have the same direction, then speed increases. When acceleration and velocity vectors oppose each other, speed decreases. When solving horizontal-motion problems, the initial motion can always be set as rightward ($+v_{xi}$).

However, vertical motion is subject to the constant downward acceleration of gravity and is more complicated. Upwardly launched objects always solve traditionally. The initial upward vertical velocity is positive ($+v_{yi}$), and the acceleration of gravity is negative ($-g$). The opposing acceleration acts to slow objects as they ascend. After reaching maximum height, objects reverse direction ($-v_y$). Now velocity and acceleration have the same direction, causing speed to increase as objects descend. When objects are dropped or thrown downward, the signs on all nonzero kinematic variables are negative. Since all the negative signs cancel, they are often omitted when solving these problems. The solutions appear as though the initial downward motion was set as the positive, resulting in positive downward velocity ($+v_y$), displacement ($+\Delta y$), and gravity ($+g$). Setting the direction of initial motion as positive simplifies the sign on acceleration. If this is done, then positive acceleration increases speed, and negative acceleration decreases speed.

The most important aspect to determining vector direction is consistency. Which direction you set as positive does not really matter. However, you must consistently apply this decision to every vector throughout the entire problem. When you pick a positive and a negative direction, do not change it during the problem. Make sure every vector uses the sign convention you have chosen.

Kinematic Equations

The kinematic equations relate the kinematic variables in a manner that solves for a variety of situations.

$$\Delta x = v_i t + \frac{1}{2}at^2 \qquad v_f = v_i + at \qquad v_f^2 = v_i^2 + 2a\Delta x$$

Choosing the Correct Equation

Choosing the correct equation depends on the variables mentioned in each problem. In addition, when an object is initially at rest ($v_i = 0$), the equations simplify into frequently tested, easier versions of the kinematic equations. Table 3.4 will help you identify which equation you should use based on what is given and what is requested in a particular problem. It will also help you identify shortened variations of those equations for objects that are initially at rest.

TABLE 3.4 Choosing Which Kinematic Equation to Use

Problem Includes	Equation	Initially at Rest ($v_i = 0$)
Time is *not* given and is *not* requested	$v_f^2 = v_i^2 + 2ax$	$v_f = \sqrt{2ax}$
Time is given and velocity is requested	$v_f = v_i + at$	$v_f = at$
Time is given and displacement is requested	$x = v_i t + \frac{1}{2}at^2$	$x = \frac{1}{2}at^2$
Constant velocity ($a = 0$)	$v = \frac{x}{t}$ or $x = vt$	Not applicable

Note: The constant velocity ($a = 0$) formula can be derived from $x = v_i t + \frac{1}{2}at^2$ by substituting zero for acceleration: $x = v_i t$. If velocity is constant, the initial velocity, v_i, is the same as the velocity, v, at any instant.

EXAMPLE 3.1

Problem Never Mentions Time

Determine the maximum height reached by a ball thrown upward at 20 meters per second.

WHAT'S THE TRICK?

Complete a variable list, including known constants and hidden values. In vertical-motion problems, y and h are often used in place of displacement, x. In addition, the acceleration of gravity, g, replaces the general acceleration, a. When objects reach "maximum height," they come to an instantaneous stop ($v_f = 0$ m/s).

$y = ?$	$v_i = 20$ m/s	$v_f = 0$ m/s	$g = -10$ m/s^2	$t = $ not mentioned

Time is not given, and you are not asked to solve for it.

$$v_f^2 = v_i^2 + 2gy$$
$$(0)^2 = (20 \text{ m/s})^2 + (2)(-10 \text{ m/s}^2)\, y$$
$$y = 20 \text{ m}$$

EXAMPLE 3.2

Problem Involves Time and Velocity

A car traveling at 30 meters per second undergoes an acceleration of 5.0 meters per second squared for 3.0 seconds. Determine the final velocity of the car.

WHAT'S THE TRICK?

Complete a variable list. If variables seem to be missing, read the problem again and look for key phrases signaling hidden variables. The problem did not state how the acceleration was affecting the car, so you must assume the simplest scenario. Unless the problem specifies a decrease in speed, assume acceleration is positive and that it acts to increase speed.

$x = $ not mentioned	$v_i = 30$ m/s	$v_f = $ determine	$a = 5.0$ m/s^2	$t = 3.0$ s

Time is not given, and the problem involves velocity.

$$v_f = v_i + at$$
$$v_f = (30 \text{ m/s}) + (5 \text{ m/s}^2)(3.0 \text{ s})$$
$$v_f = 45 \text{ m/s}$$

EXAMPLE 3.3

Problem Involves Time and Displacement

A ball is dropped from a 45-meter-tall structure. Determine the time the ball takes to hit the ground.

WHAT'S THE TRICK?

Complete a variable list, including known constants and hidden values. A "dropped" object has an initial velocity of zero ($v_i = 0$ m/s). The structure is 45 m tall, and the ball is moving downward toward the ground ($y = -45$ m). The acceleration is due to gravity, which also acts downward ($g = -10$ m/s^2).

$y = -45$ m	$v_i = 0$ m/s	$v_f =$ not mentioned	$g = -10$ m/s^2	$t =$ determine

You need an equation relating displacement and time.

$$y = v_i t + \frac{1}{2}gt^2$$

Since $v_i = 0$ m/s, you can simplify the equation.

$$y = \frac{1}{2}gt^2$$
$$(-45 \text{ m}) = \frac{1}{2}(-10 \text{ m/s}^2)t^2$$
$$t = 3.0 \text{ s}$$

Kinematic Graphs

The key values to assess are slope, area, and intercepts. To determine if slope or area is important, remember to include units in your calculations. In addition, it may also be important to determine if values are constant or changing. Table 3.5 describes frequently used graphs involving the kinematic formulas and variables.

TABLE 3.5 Graphs and Kinematics

If You See . . .	Slope	Area	y-intercept	x-intercept
Position, x (or distance, d) versus time	Velocity		Initial position	$x = 0$
Velocity (or speed) versus time	Acceleration	Displacement (change in position)	Initial velocity	$v = 0$
Acceleration versus time		Change in velocity	Initial acceleration	$a = 0$

The velocity versus time graph described in Table 3.5 contains the most information, making it the most valuable and most frequently tested kinematic graph.

EXAMPLE 3.4

Analyzing Velocity versus Time Graphs

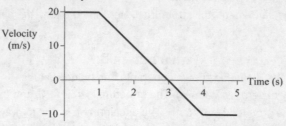

The motion of an object is shown in the velocity versus time graph above.

(A) Determine the initial velocity of the object.

WHAT'S THE TRICK?

Initial conditions occur at zero time. In graphs with time along the *x*-axis, initial values are the *y*-intercept. The initial velocity is 20 m/s.

(B) Determine the displacement during the first second.

WHAT'S THE TRICK?

Displacement is the area under the velocity versus time graph.

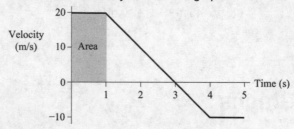

$$\text{displacement} = \text{height} \times \text{base} = (20 \text{ m/s} \times 1 \text{ s}) = 20 \text{ m}$$

(C) Determine the acceleration in the time interval between 1 and 4 seconds.

WHAT'S THE TRICK?

Acceleration is the slope of the velocity versus time graph.

$$\text{slope} = \frac{\text{rise}}{\text{run}} = \frac{(-10 \text{ m}) - (20 \text{ m})}{(4s) - (1s)} = -10 \text{ m/s}^2$$

CHAPTER 4

Kinematics in Two Dimensions

Learning Objectives

In this chapter, you will learn how to:

o Define and discuss the independence of motion

o Determine the true velocity and displacement of objects

o Solve projectile-motion problems

Table 4.1 lists the variables that will be used and their units.

TABLE 4.1 Variables Used in Two-Dimensional Kinematics

New Variables	Units
$\vec{v}_i$ = True, three-dimensional initial velocity	m/s (meters per second)
$\vec{v}_f$ = True, three-dimensional final velocity	m/s (meters per second)
$\vec{s}$ = True, three-dimensional displacement	m (meters)
v_{ix} = Initial velocity in the x-direction	m/s (meters per second)
v_{iy} = Initial velocity in the y-direction	m/s (meters per second)
v_{fx} = Final velocity in the x-direction	m/s (meters per second)
v_{fy} = Final velocity in the y-direction	m/s (meters per second)

Independence of Motion

You already know how kinematic equations are used to determine the position, velocity, and acceleration of an object moving along a one-dimensional line. Consider an example of an astronaut in space throwing a ball horizontally with an initial velocity of $\vec{v}_i$. If the astronaut is very far from Earth, and gravity is negligibly small, the ball will continue to move in a straight line. The diagram below indicates the instantaneous velocity vectors on the ball at four different locations during its motion.

$$\vec{v}_i \rightarrow \qquad v_1 = \vec{v}_i \rightarrow \qquad v_2 = \vec{v}_i \rightarrow \qquad v_3 = \vec{v}_i \rightarrow$$

FIGURE 4.1 Velocity vectors

Notice that all the velocity vectors have equal magnitudes and directions. If no external forces are acting on the ball after its release, the ball will continue moving with its initial velocity, $\vec{v}_i$, indefinitely.

Consider if someone throws the same ball horizontally near the surface of Earth. If the ball is given the same initial velocity, $\vec{v}_i$, its path will resemble a parabola.

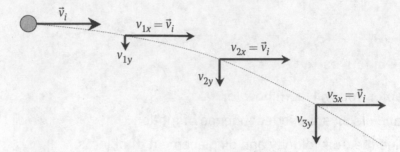

FIGURE 4.2 Projectile velocity vectors

The ball will continue to move in the x-direction at velocity $\vec{v}_i$. However, it will also experience a vertical acceleration due to gravity. As a result, the vertical velocity, v_y, will increase uniformly. The resulting path of the ball is a parabola.

Kinematic equations can be used for motion only along a straight line. Therefore, separate kinematic equations must be employed for x-variables and for y-variables. The resulting motion is described by two kinematic equations in combination. Adding x- and y-subscripts to the kinematic variables allows you to distinguish between similar variables acting in different directions. Table 4.2 compares the kinematic equations in one and in two dimensions.

TABLE 4.2 Kinematic Equations in One and Two Dimensions

One-dimensional Kinematics Equations	Two-dimensional Kinematic Equations	
	x-direction	y-direction
$x = v_i t + \frac{1}{2} a t^2$	$x = v_{ix} t + \frac{1}{2} a_x t^2$	$y = v_{iy} t + \frac{1}{2} a_y t^2$
$v_f^2 = v_i^2 + 2ax$	$v_{fx}^2 = v_{ix}^2 + 2a_x x$	$v_{fy}^2 = v_{iy}^2 + 2a_y y$
$v_f = v_i + at$	$v_{fx} = v_{ix} + a_x t$	$v_{fy} = v_{iy} + a_y t$

Although the kinematic equations can never contain a mixture of x- and y-variables, the individual x- and y-equations do share one very important variable—time, t. The mathematically independent x- and y-motions take place simultaneously and share the same time, t.

True Velocity and Displacement

The kinematic equations solve for x- and y-direction velocities and displacements. However, in two-dimensional motion problems, the path followed by the object does not lie solely along either the x- or y-axis. The ball in Figure 4.3 follows a parabolic path.

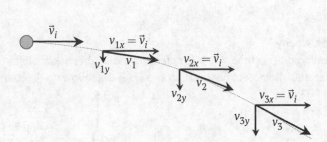

FIGURE 4.3 Projectile motion velocity vectors and their components

At any instant during this motion, there is the **true velocity**, $\vec{v}$, tangent to the motion of the object. The true velocity is found using vector mathematics. The x- and y-velocities, calculated by the separate one-dimensional kinematic equations, are the components of the true velocity. You can find the true velocity of the object by using the Pythagorean theorem.

$$|\vec{v}| = \sqrt{v_x^2 + v_y^2}$$

The **true displacement** of an object can be found in a similar manner. The kinematic equations solve separately for the x- and y-displacements. These displacements are also the vector components of the true displacement of an object. Use the Pythagorean theorem to find the true displacement.

$$|\vec{s}| = \sqrt{x^2 + y^2}$$

Relative Velocity

The motion of an object, as described by two observers, may differ depending on the location of the observers. For example, a car reported as moving at 30 meters per second by a stationary observer will appear to be moving at 10 meters per second as observed by a driver in a car traveling alongside at 20 meters per second. Problems involving multiple velocities are known as **relative-velocity** problems. Common two-dimensional relative-velocity problems involve a boat moving across a river or an airplane flying through the air. In these problems, the velocities in both the x- and y-directions are constant. Therefore, the acceleration in the x-direction, a_x, and the acceleration in the y-direction, a_y, are both equal to zero, as shown in Table 4.3. This greatly simplifies the kinematic equations.

TABLE 4.3 Relative-Velocity Equations

Kinematic Equation Used for Relative Velocity	Modified for x-direction When $a_x = 0$	Modified for y-direction When $a_y = 0$
$x = v_i t + \frac{1}{2} a t^2$	$x = v_{ix} t$	$y = v_{iy} t$

The velocities of river currents and the wind change the true velocity of boats and airplanes. Relative-velocity problems require you to understand both vector addition and independence of motion.

EXAMPLE 4.1

Determining True Velocity

An airplane is heading due north at 400 kilometers per hour when it encounters a wind from the west moving at 300 kilometers per hour, as shown in the following diagram.

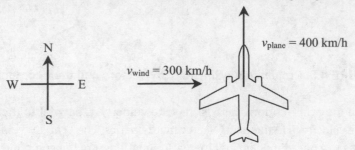

(A) Determine the magnitude of the true velocity of the plane with respect to an observer on the ground.

WHAT'S THE TRICK?

An observer on the ground will see the true velocity of the airplane. Mathematically, this is the resultant vector created by adding the airplane and wind velocity vectors tip to tail, as shown below, and using the Pythagorean theorem.

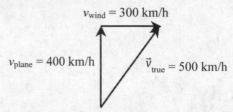

The airplane has a speed of 500 kilometers per hour relative to the ground.

(B) Which vector, of the choices given below, describes the direction the pilot must aim the plane in order for the plane to have a true velocity that points directly north?

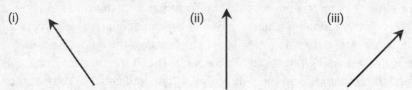

(i)   (ii)   (iii)

WHAT'S THE TRICK?

The velocity vectors for the plane and the wind must add together to create a true velocity that points north. Since the wind is blowing out of the west, the plane must have a component that moves toward the west to cancel the effect of the wind.

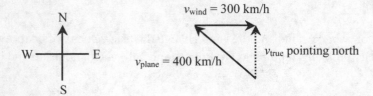

Answer (i) is the correct heading for the plane. When the plane is aimed northwest, it will actually move with a true velocity directly north.

Projectile Motion

Projectile motion describes an object that is thrown, or shot, in the presence of a gravity field. An object is considered a projectile only when it is no longer in contact with the person or device that has thrown it and before it has come into contact with any surfaces. As a result, the downward acceleration of gravity is the only acceleration acting on a projectile during its flight.

One of the most important aspects of projectile motion is the type of motion experienced in each direction. If the vertical acceleration of gravity is the only acceleration acting on a projectile, the horizontal speed of a projectile cannot change. Therefore, the horizontal component of velocity must always remain constant. Both the vertical velocity and the vertical displacement will be affected by the acceleration of gravity. When calculating the vertical portion of projectile motion, you must use the complete kinematic equations, as shown in Table 4.4. You should include additional subscripts to distinguish the x- and y-velocities from each other and from the true velocity, $\vec{v}$.

TABLE 4.4 Kinematic Equations with Gravity

x-direction Constant Velocity: $a_x = 0$	y-direction Gravity: $a_y = g$	True Velocity at Any Instant
$x = v_{ix}t$ $v_{fy} = v_{iy}$	$y = v_{iy}t + \frac{1}{2}gt^2$ $v_{fy}^2 = v_{iy}^2 + 2gy$ $v_{fy} = v_{iy} + gt$	$|\vec{v}| = \sqrt{v_x^2 + v_y^2}$

The first step in projectile-motion problems involves determining the initial velocity in the x-direction, v_{ix}, and the initial velocity in the y-direction, v_{iy}. These velocities are the components of the initial launch velocity, $\vec{v}_i$. Table 4.5 shows the velocities of two typical launches.

TABLE 4.5 Velocities of Two Typical Launches

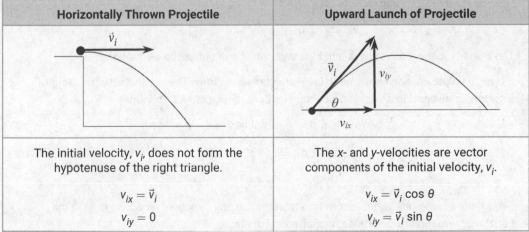

Horizontally Thrown Projectile	Upward Launch of Projectile
The initial velocity, v_i, does not form the hypotenuse of the right triangle.	The x- and y-velocities are vector components of the initial velocity, v_i.
$v_{ix} = \vec{v}_i$ $v_{iy} = 0$	$v_{ix} = \vec{v}_i \cos\theta$ $v_{iy} = \vec{v}_i \sin\theta$

The next step in projectile motion usually involves determining time, t. As with all two-dimensional motion problems, time is the only variable common to both the x- and y-directions. You should note that the time of flight, t, depends on y-direction variables, not on x-direction variables. Therefore, when the time of flight is not given, most problems begin by solving one of the three y-direction kinematic equations.

Horizontally Launched Projectiles

Horizontally launched projectiles are the most common projectile-motion problems encountered on introductory physics exams. As with any kinematics problem, identifying variables (especially hidden variables) is extremely important. In horizontal launches, the initial velocity in the y-direction is zero, as shown in Table 4.6. This simplifies the y-direction equations.

TABLE 4.6 Kinematic Equations for Horizontal Launches

	x-equation	Simplified y-equations
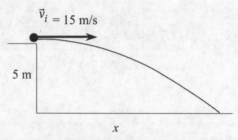	$x = v_{ix}\, t$	$y = \frac{1}{2} g t^2$ $v_{fy} = \sqrt{2gy}$ $v_{fy} = gt$

EXAMPLE 4.2

Horizontally Launched Projectiles

$$\vec{v}_i = 15 \text{ m/s}$$

5 m

x

A ball is thrown horizontally at 15 meters per second from the top of a 5-meter-tall platform, as shown in the diagram above. Determine the horizontal distance traveled by the ball.

WHAT'S THE TRICK?

First determine the x- and y-components of the initial velocity. This is easy for horizontal launches. All of the initial velocity is directed horizontally, and none is directed vertically.

$$v_{ix} = \vec{v}_i = 15 \text{ m/s} \qquad \text{and} \qquad v_{iy} = 0$$

The velocity in the y-direction is a hidden zero, which simplifies the y-equations.

If time is unknown, solve for time using y-direction equations. The vertical displacement of 5 meters is given. Use an equation containing both displacement and time.

$$y = v_{iy}t + \frac{1}{2}gt^2 \qquad \text{simplifies to} \qquad y = \frac{1}{2}gt^2$$

$$(-5 \text{ m}) = \frac{1}{2}(-10 \text{ m/s}^2)\, t^2$$

$$t = 1 \text{ s}$$

Finally, time is the one variable common to motion in any direction. Now, use time in the horizontal equation to determine the horizontal distance.

$$x = v_{ix}t$$

$$x = (15 \text{ m/s})(1 \text{ s}) = 15 \text{ m}$$

Projectiles Launched at an Angle

Projectiles launched at angles are more difficult to solve mathematically. Figure 4.4 depicts the flight path of a projectile launched with an initial speed of $\vec{v}_i = 50$ m/s at an upward launch angle. The projectile lands at the same height from which it was launched, $y = 0$, and has a final speed $v_f = 50$ m/s. The projectile is shown every second during its flight. In each of these positions, the instantaneous horizontal component of velocity, v_x, and the instantaneous vertical component of velocity, v_y, are shown.

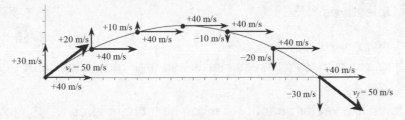

FIGURE 4.4 Projectile-motion vectors

Examination of the diagram reveals four key facts about projectiles launched at angles.

1. The horizontal component of velocity, v_x, remains constant.

2. When the projectile is moving upward, its vertical speed decreases by 10 meters per second every second until the projectile reaches an instantaneous vertical speed of zero at maximum height. The decreasing velocity then results in a changing downward speed that increases by 10 meters per second every second.

3. The projectile passes through each height twice, once on the way up and once on the way down (except the single point at maximum height). At points with equal height, the magnitude of the vertical velocity on the way up equals the magnitude of the vertical velocity on the way down.

4. The time the projectile rises equals the time the projectile falls, as long as the final height equals the initial height.

Retaining a mental image of the above diagram in your memory will help you answer conceptual problems for upwardly launched projectiles. The most common problems tend to focus on two key locations during the flight: the very top of the flight path (maximum height) and the landing point, as shown in Figure 4.5.

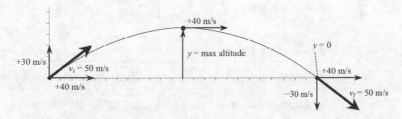

FIGURE 4.5 Two key instants during a projectile flight

CHAPTER 5

Dynamics

Learning Objectives

In this chapter, you will learn:

o Why objects with mass possess inertia and their natural state of motion is constant velocity

o Why a force is a vector quantity that can be thought of as a push or a pull acting to change the motion of an object

o The three laws of motion postulated by Sir Isaac Newton summarizing the interactions among objects, forces, and motion

o How their influence on kinematics will be encountered in all areas of physics, developing universal problem-solving strategies for all dynamics problems, an essential skill

Table 5.1 lists the variables used in dynamics.

TABLE 5.1 Variables and Units for Forces in One Dimension

New Variables	Units
$\vec{F}$ = Applied force	N (newtons)
$\vec{F}_g$ = Force of gravity or weight	
$\vec{N}$ = Normal force	
$\vec{f}_s$ = Static friction	
$\vec{f}_k$ = Kinetic friction	
$\vec{T}$ = Tension	
$\vec{F}_s$ = Force of elastic devices, springs	
k = Spring constant	N/m (newtons per meter)
$\Sigma\vec{F}$ = Net force or sum of forces	N (newtons)
ΣF_x = Net force along the x-axis	
ΣF_y = Net force along the y-axis	

Inertia

Inertia is the tendency of an object to resist changes in its natural motion. Galileo suggested that the natural state of motion of an object is constant velocity. Even an object at rest has a constant velocity of zero. Inertia is the tendency of a stationary object to remain at rest and for a moving object to continue moving at constant velocity. Inertia is a property of mass. The greater the mass of an object, the greater the inertia possessed by the object. Simply put, the more mass the object has, the harder it is to push around the object. Physicists regard mass as a quantity of inertia and a resistance to change in motion.

Force

An object cannot simply accelerate and change velocity on its own. In order to accelerate, an object must be pushed or pulled by some external source known as an **agent**. The push or pull of an agent that accelerates an object is known as a **force**. Forces are vector quantities measured in newtons.

Since forces are vector quantities, they can be added together to find a resultant sum. The sum of all forces, $\Sigma \vec{F}$, acting on an object is known as the **net force**. The direction of the net force is the same as the direction of the acceleration (change in velocity) of the object. The relationship between the direction of force and the direction of initial velocity of an object dictates the resulting change in motion, as shown in Table 5.2.

TABLE 5.2 Relationship Between Direction of Force and Direction of Initial Velocity

Direction of Force Matches Initial Motion	Direction of Force Opposes Initial Motion	Direction of Force Perpendicular to Initial Motion
$\vec{v}$ m $\vec{F}$	$\vec{v}$ $\vec{F}$ m	$\vec{v}$ m $\vec{F}$
Forces acting in the direction of motion attempt to increase the speed of objects. These forces can be considered positive.	Forces acting opposite the direction of motion attempt to decrease the speed of objects. These forces can be considered negative.	The force will cause the object to change direction.

Common Forces

Numerous agents, each with unique characteristics, are capable of generating forces. As a result, several important forces have been given unique variables.

Applied Force

The general variable letter, F, is used to represent any force that does not have its own variable designation. As an example, there is no specific variable for the force of a person pushing a box.

Force of Gravity (Weight)

Objects having mass are considered to be surrounded by a mathematical vector field known as the gravitational field, g. The gravity field is essentially the value of the acceleration of gravity at every point in space surrounding a mass. All masses are surrounded by a gravity field. However, when two masses interact, one of the masses (usually the larger mass) is thought of as the agent surrounded by a gravity field. The second mass (usually the smaller mass) is thought of as the object. When the second mass (object) is placed in the gravity field of the first mass (agent), the magnitude of the force of attraction on the object is the product of the object's mass, m, and the agent's gravity field, g.

$$F_g = mg$$

This scenario can also be reversed. The force of gravity acting on the agent due to the gravity field surrounding the object has the same magnitude, but acts in the opposite direction according to Newton's third law.

Most problems take place on Earth. Unless stated otherwise, Earth is the agent creating the gravity field, and $g = 10 \text{ m/s}^2$. As shown in Figure 5.1, the force of gravity pulls the object with mass m toward the agent, Earth.

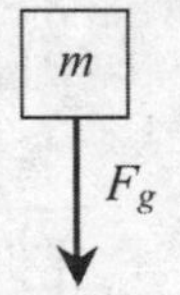

FIGURE 5.1 Force of gravity (weight)

The force of gravity is also the **weight** of an object. Some instructors, texts, and exams may use the variable w to represent weight (force of gravity).

$$w = F_g = mg$$

Normal Force

The **normal force**, N, is present whenever an object pushes on a surface. There is no specific formula to solve for the normal force. Rather, the normal force is a response force to an applied force. As will be shown shortly, the normal force is a result of Newton's third law. As shown in Figure 5.2, the normal force always acts perpendicular to a surface.

FIGURE 5.2 Normal force

Friction

The **force of friction**, f or F_f, is present when two conditions are met. First, an object must be pressed against a rough surface. Second, either a force or a component of force must be acting parallel to the surface. Figure 5.3 shows the applied force, F, the force due to friction, f, the force due to gravity, F_g, and the normal force, N, acting on an object with mass m.

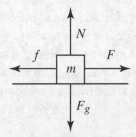

FIGURE 5.3 Free-body diagram

In Figure 5.3, the force of gravity, F_g, pulls the object onto the surface. This causes the surface to press back with the normal force N. When any force, in this case force F, is applied parallel to a surface, that force will attempt to accelerate the object. If the surface is rough, then a frictional force, f, will act opposite the applied force.

Friction is described by the following equation and is dependent on two quantities:

$$f \leq \mu N$$

1. The normal force, N, is the responding support force of a surface that results when an object presses against the surface. Note that surface area is not present in the above formula. Friction does not depend on surface area. For an object with constant mass, surface area is inversely proportional to the pressure acting per unit of surface area. These factors completely offset each other. As a result, rotating an object to a different surface with a different surface area will not change the magnitude of the normal force nor the magnitude of friction.

2. The coefficient of friction, μ (no units), is a ratio of the force of friction to the normal force. There are two categories of friction. When friction acts and an object remains stationary, it is subject to **static friction**, f_s, which involves the **coefficient of static friction**, μ_s. The friction acting on moving objects is known as **kinetic friction** and involves the **coefficient of kinetic friction**, μ_k. The coefficients of static and kinetic friction will have two different values for the same object, and the coefficient of static friction is greater than the coefficient of kinetic friction.

$$\mu_s > \mu_k$$

When an object is initially at rest on a rough surface and a force, or component of force, is applied parallel to the surface, an opposing friction force responds simultaneously. If the object remains at rest, then it is subject to static friction. There are two ways to solve for the magnitude of static friction. The first method is the most advantageous as it works consistently. When an object remains at rest, it is in static equilibrium, and the forces acting on it must be equal and opposite. Therefore, the magnitude of static friction will be equal to the magnitude of the opposite applied force acting parallel to the surface.

$$f_s = F_{\text{applied}}$$

The second method involves the friction equation ($f \leq \mu N$). This is problematic due to the "less than or equal to" mathematical operator. When the equation is set as an equality, then it will only solve for the **maximum value of static friction**, $f_{s\,\text{max}}$.

$$f_{s\,\text{max}} = \mu_s N$$

However, static friction can be less than the value of maximum static friction. In static problems, the friction equation can only be used when a problem indicates that static friction has reached its limit, and that even a slight increase in the forward applied force would cause the object to slip. To avoid mistakes in static problems, start by attempting the first method, which works consistently. If this does not arrive at an answer, then use the friction equation. When objects experience maximum static friction, solutions may require the use of both static friction equations.

Kinetic friction can act on objects that accelerate or that move at constant velocity. Fortunately, the friction equation always solves as an equality for moving objects.

$$f_k = \mu_k N$$

Kinetic-friction problems involving acceleration can only use the above equation. However, when an object is moving at constant velocity, the forces acting parallel to motion are equal and opposite. Therefore, kinetic-friction problems involving constant velocity can also be solved by setting kinetic friction equal to the applied forward force.

$$f_k = F_{\text{applied}}$$

When objects move at constant velocity, solutions may require the use of both kinetic friction equations.

Tension

Tension, T, is a force exerted by ropes and strings. There is no specific formula for tension. As with the normal force, tension is a response force and obeys Newton's third law. Tension acts along ropes or strings as shown in Figure 5.4.

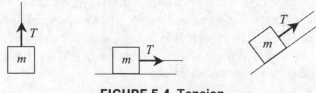

FIGURE 5.4 Tension

Tension has the same magnitude in every part of a particular rope or string.

Springs

When springs are stretched or compressed by an outside agent, a force is created in the spring. The force of a spring is known as a **restoring force**, and it acts to restore the spring to its original rest length. As a result, the force of a spring, F_s, always opposes the action of the agent. If an agent stretches a spring, the restoring force acts to compress the spring. If an agent compresses a spring, the restoring force acts to stretch the spring. See Figure 5.5.

FIGURE 5.5 The force of a spring

The magnitude of the force of a spring is described in **Hooke's law**:

$$F_s = |kx|$$

The variable k is the **spring constant** (units: N/m), and the variable x is the distance the spring is either stretched or compressed (units: m). Every spring has its own unique spring constant. Finding the spring constant, if it is not given, is often the first critical step in solving spring problems.

Force Diagrams

The first step in solving a dynamics problem is to identify all forces, including the direction of each force, acting on an object. This can be accomplished by constructing a force diagram, known as a **free-body diagram**. A free-body diagram includes only the object and the forces acting on the object. It is free of clutter and does not include surfaces, strings, springs, or any other agents acting on the object. A free-body diagram serves two key purposes: First, it allows you to visualize how the forces will influence the motion. Second, it provides a frame of reference in which to work with the force vectors.

EXAMPLE 5.1

Free-Body Diagrams

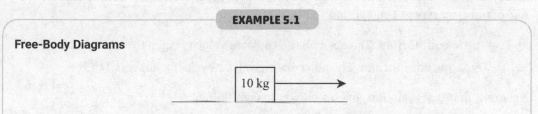

A 10-kilogram mass is pulled at constant velocity to the right along a rough horizontal surface by a string. Construct a free-body diagram depicting all the forces acting on the mass.

WHAT'S THE TRICK?

Identify all the forces acting on the mass.

- Gravity is not mentioned in the problem. Unless specified otherwise, problems take place on Earth, so the force of gravity, F_g, is present.
- You may be able to deduce some forces from the diagram. Both a string and a surface are in contact with the object. Therefore, tension, T, and the normal force, N, are present.
- Identifying other forces requires you to read the text of the problem. This problem mentions a "rough" surface, implying that friction, f, is present.

The resulting free-body diagram is shown below left. Free-body diagrams are nearly identical to plotting vectors on coordinate axes, as shown below right.

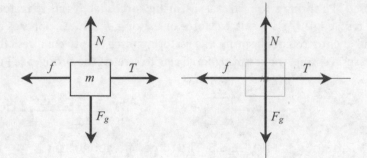

Including all force vectors and ensuring that they point in the correct direction are the most important aspects when drawing a free-body diagram.

Newton's Laws of Motion

Sir Isaac Newton deduced three laws of motion to describe the relationships among force, mass, and changes in velocity.

Newton's First Law

In his first law, Newton simply generalizes Galileo's principle of inertia. Essentially, the first law states that objects at rest will remain at rest, and objects in motion will remain at constant velocity unless acted upon by a net unbalanced force. This law is often referred to as the **law of inertia**.

When the word *equilibrium* is used in physics, the forces acting on an object are balanced (equal and opposite). Equilibrium is a stable condition where the net force is zero, $\Sigma \vec{F} = 0$, and an object does not accelerate, $a = 0$. If the object is not accelerating, it must be experiencing one of only two types of equilibrium.

1. **Static equilibrium.** The object has a constant velocity equal to zero.
2. **Dynamic equilibrium.** The object has a constant velocity not equal to zero.

Newton's first law of motion governs objects in equilibrium.

Newton's Second Law

Newton's second law addresses changes to the inertia of an object. It states that when an object of mass m is acted upon by a net force of $\vec{F}_{net}$ or $\Sigma \vec{F}$ (the sum of all forces acting together), the object will accelerate. In addition,

- the acceleration of the object will be in the direction of the net force,
- the acceleration will be directly proportional to the net force, and
- the acceleration will be inversely proportional to the mass of the object.

These details are very concisely summarized in one the most famous equations in physics.

$$\Sigma\vec{F} = m\vec{a}$$

Newton's second law results in several important consquences.

1. When all the one-dimensional vector forces sum to a nonzero value, the forces are said to be unbalanced. The result is acceleration, which alters either the magnitude or the direction of velocity. When an object is accelerating, there must be a net, unbalanced force acting on the object.

2. The direction of acceleration is the same as the direction of the net force.

3. The direction of velocity is not necessarily in the same direction as the acceleration. An example would be the change in velocity of a car as it slows to a stop. The initial velocity is in one direction, but the acceleration is in the opposite direction, causing the car to slow down.

4. When vector forces add up to zero, the forces are balanced. There can be no acceleration. An object will continue at constant velocity based on the first law of motion. When an object is either at rest or moving at constant velocity, the one-dimensional forces acting on the object must cancel each other. The net force must equal zero.

Newton's Third Law

Newton's third law states that when objects interact, equal and opposite forces act simultaneously on both objects. Most problems focus only on the forces that act to determine the motion of a specific mass referred to as the "object." However, objects cannot exert a net force on themselves, which means they cannot accelerate themselves and cannot change their own velocities. Objects must be pushed or pulled by another entity known as the "agent." The forces that act on the object causing it to move are created by the agent. According to Newton's third law, when an agent pushes an object with an **action force**, the object simultaneously pushes back with an equal and opposite **reaction force**. This can also be viewed from a reversed frame of reference. When an object pushes on the agent with an **action force**, the agent simultaneously pushes back on the object with an equal and opposite **reaction force**. These simultaneous forces are referred to as an action-reaction pair. As an example, a person walking does not directly use their leg muscles to push themselves forward. They actually use their leg muscles to push their feet backward along the surface of the Earth (action), and simultaneously the surface of Earth pushes back with an equal and opposite friction force (reaction) that propels the person forward. While the problem may only focus on the force acting to move the person forward, this force cannot occur by itself. If you do not push your feet backward, the earth cannot push you forward, and if the surface of the earth were frictionless, no movement of the feet or legs would result in forward motion. Newton's third law is useful in determining a variety of forces such as the normal force, tension, and friction. It is also very evident in problems involving collisions and recoil, as seen in explosions. While a problem may only seem to focus on the forces acting on the object, it may also be important to consider the effect and consequences due to the equal and opposite force acting on the agent.

Solving Force Problems

The following four-step method is a suggested attack plan that breaks difficult force problems into a series of steps you can easily solve.

1. **Orient the problem.** Identify force vectors and the relevant directions. Then sketch a force diagram.

2. **Determine the type of motion.** Determine if the problem involves
   - equilibrium, constant velocity, balanced forces, $\Sigma \vec{F} = 0$ or
   - dynamics, acceleration, unbalanced forces, $\Sigma \vec{F} = m\vec{a}$.

3. **Sum the force vectors in the relevant direction.** Only forces, or components of forces, parallel to the direction of motion can change the speed of an object. Sum only the one-dimensional force vectors parallel to the object's motion. Any force vectors pointing in the direction of motion will increase the speed of the object and should be set as positive. Force vectors opposing the motion of the object should be set as negative.

4. **Substitute and solve.** In this last step, substitute known equations for specific forces and then substitute numerical values to find the solution.

The following sections include common example problems solved using this four-step strategy.

Inclines

Incline problems are a special case where the x- and y-axes are not useful. Instead, it is easier to work with tilted axes that are parallel and perpendicular to the incline, as shown in Figure 5.6(a). The left diagram also reveals that although the normal force lies on one of axes, the force of gravity does not.

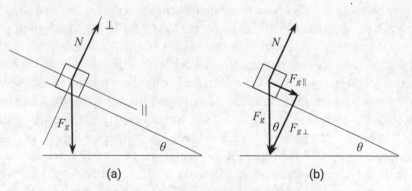

FIGURE 5.6 Solving incline problems

As a result, the force of gravity must be split into components, as shown in Figure 5.6(b). Solving the vector components of the force of gravity results in the following equations:

- Force of gravity parallel to the incline: $F_{g\parallel} = mg \sin \theta$
- Force of gravity perpendicular to the incline: $F_{g\perp} = mg \cos \theta$

The force of gravity perpendicular to the incline pushes the object into the incline. The incline pushes back with an equal and opposite force, creating the normal force, N.

$$N_{\text{incline}} = F_{g\perp} = mg\cos\theta$$

The component of the force of gravity parallel to the incline is unbalanced. If no other forces are present, this component will cause the object to accelerate down the incline. Whenever an incline is present, there is always a component of force acting to pull an object down and parallel to the incline.

$$F_{g\parallel} = mg\sin\theta$$

The relationship and difference between the force of gravity, F_g, and the force of gravity parallel to the incline, $F_{g\parallel}$, confuses many students. The force of gravity is the actual force acting on the object and is always included in free-body diagrams. However, it is not used to solve an incline problem mathematically. Calculating the mathematical solution for motion parallel to an incline requires force vectors or components of force vectors that are parallel to the incline. The force of gravity parallel to the incline, $F_{g\parallel}$, is a component of force that is always present in an incline problem. Since it is a component of a force vector, $F_{g\parallel}$ is never included in the free-body diagram. However, it is always included in calculations of the net force acting parallel to an incline. Figure 5.7(a) shows the free-body diagram on an incline. Figure 5.7(b) shows the vector component needed when solving equations for the same situation.

(a)   (b)

FIGURE 5.7 Force on an incline

EXAMPLE 5.2

Inclines

A 5.0-kilogram mass is positioned on a 30° frictionless incline. It is kept stationary by a string pulling parallel to the incline. Determine the tension in the string.

WHAT'S THE TRICK?

Work with forces and components that are parallel and perpendicular to the slope.

Orient the problem: The force diagram below left shows all forces and the components of the force of gravity parallel and perpendicular to the incline. However, to find the tension T, only forces parallel to the slope are needed. The diagram below right shows only the forces relevant to this problem.

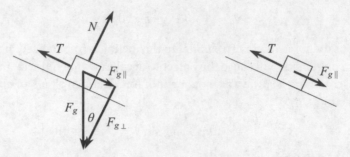

Note: The components of F_g are *not* included in a formal free-body diagram.

Determine the type of motion: The problem asks for the tension in the string that will keep the mass at rest. The mass obeys Newton's first law, $\Sigma F_{||} = 0$.

Sum the force vectors in the relevant direction: There are two vectors parallel to the incline. Normally, the direction of motion is set as positive. However, the masses are not moving. So, you can set either direction as positive.

$$\Sigma F_{||} = F_{g||} - T$$

Substitute and solve: Substitute zero for $\Sigma F_{||}$, substitute $mg \sin \theta$ for $F_{g||}$, substitute numerical values, and solve.

$$0 = mg \sin \theta - T$$

$$T = mg \sin \theta$$

$$T = (5.0 \text{ kg})(10 \text{ m/s}^2) \sin 30^\circ$$

Remember: $\sin 30^\circ = 0.5$

$$T = 25 \text{ N}$$

Dynamics and Kinematics

Dynamics and kinematics are explicitly linked. Students are commonly given a set of forces and asked to determine their effect on the motion of an object. Typically, the sum of forces helps students find the acceleration. Then the acceleration is used in the kinematic equations to determine displacement and final velocity.

EXAMPLE 5.3

Combining Dynamics and Kinematics
Mass m is initially at rest and then is pushed by force F through a distance, x, reaching a final velocity of v. What will be the new final velocity for mass m if the experiment is repeated with a force of $2F$?

WHAT'S THE TRICK?
To find the effect on velocity of a changing force, first determine how acceleration is altered. The force portion of this problem involves a single force whose direction is not important. In cases such as this, the four-step method is not needed. The lone force is responsible for all the acceleration.

$$F = ma$$

Mass remains constant. As a result, force and acceleration are directly proportional. Therefore, doubling the force doubles the acceleration.

$$(2F) = m(2a)$$

The rest of this problem involves analyzing kinematics. This problem has two important aspects. First, there is no mention of time. Second, the object starts at rest.

$$v_f^2 = v_i^2 + 2ax \quad \text{and} \quad v_i = 0$$
$$v_f = \sqrt{2ax}$$

From above, doubling the force doubles the acceleration.

$$(\sqrt{2})v_f = \sqrt{2(2a)x}$$

Acceleration is under the square root. Doubling it is the same as multiplying the velocity by the square root of 2. The new final velocity is $\sqrt{2}v_f$.

Compound-Body Problems

A compound body consists of two or more separate masses experiencing the same motion. The masses may be pressing against one another or be tied together with strings so that they move as a single system. Figures 5.8(a) and 5.8(b) illustrate both types of problems.

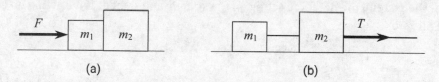

(a)   (b)

FIGURE 5.8 Compound-body problems

Both masses experience the same acceleration. The net force pushes or pulls on the total mass of the system. When solving for acceleration, you can sum the masses into a single system (sys).

$$\Sigma F_{\text{sys}} = (m_1 + m_2)a$$

In compound-body problems, there are two categories of forces to consider. **Internal forces** are the forces acting between the masses that make up the system. In Figure 5.8(a), the masses push on each other, with their surfaces creating equal and opposite internal normal forces. In Figure 5.8(b), the masses are connected with a string, and each block is pulled by an equal and opposite internal tension force. Internal forces are always equal and opposite. When the system is viewed as a single entity, the internal forces always cancel each other, and they have been left out of Figure 5.8 to simplify the diagrams. **External forces** are the forces that act on the entire system and contribute to the net force acting on all objects. In Figure 5.8(a), force F pushes both masses simultaneously, and in Figure 5.8(b), tension T pulls both masses simultaneously. A variety of variables, such as acceleration of the masses, can be solved very easily by summing the forces acting on the entire system. However, when solving for the magnitudes of the internal forces, a different strategy is needed. If asked to determine the internal forces acting between two masses, work with only one of the masses individually. Draw a free-body diagram for one mass only, and then sum the forces for that mass. The internal forces will no longer cancel. While the internal forces act in opposite directions on each mass, they always act with equal magnitudes. As a result, it does not matter which mass you choose to work with when solving for internal forces.

Compound bodies can also be oriented vertically, as shown in Figures 5.9(a) and 5.9(b). In these problems, gravity is the external force pulling the masses downward.

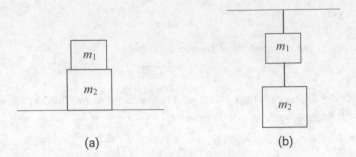

(a)   (b)

FIGURE 5.9 Vertical compound bodies

If any horizontal external forces act on either mass in Figure 5.9(a), then friction may be present between the two masses. However, in beginning physics courses, stacked masses most often remain stationary. When the systems shown in Figure 5.9 remain stationary, the acceleration of the system, the net force acting on the system, and the net force acting on each mass are all zero.

$$a = 0 \qquad \Sigma F_{sys} = 0 \qquad \Sigma F_1 = 0 \qquad \Sigma F_2 = 0$$

If the sum of forces and acceleration are zero, the forces acting on each mass must be balanced. The normal forces in Figure 5.9(a) must be equal and opposite the force of gravity pulling down against each surface. The magnitude of the normal force will be equal to but opposite the force of gravity of all the masses stacked above a surface. In Figure 5.9(b), the tension vectors are determined in a similar manner. The tension in a string is equal to but opposite the force of gravity for all the masses hanging from a string.

EXAMPLE 5.4

Compound Bodies

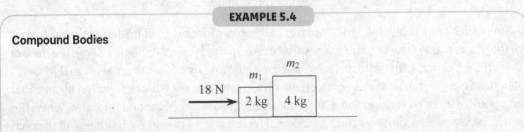

A force of 18 newtons pushes two masses ($m_1 = 2$ kilograms and $m_2 = 4$ kilograms) horizontally to the right.

(A) Determine the acceleration of mass m_2.

WHAT'S THE TRICK?

Both masses are moving together as a system. The acceleration of mass m_2 equals the acceleration of the system.

Orient the problem: Visualize the masses as a single system.

18 N → $m_1 + m_2$

Determine the type of motion: The problem asks for acceleration.

$$\Sigma F_{sys} = (m_1 + m_2)a$$

Sum the force vectors in the relevant direction: The 18-newton force in the diagram above is the only external force in the direction of motion that is not canceled. Internal normal forces act horizontally between m_1 and m_2. However, Newton's third law dictates that the forces are opposite and equal. Therefore, the forces cancel each other. In addition, any vertical forces would not affect the net force in the horizontal direction. Only the 18-newton force remains.

$$\Sigma F_{sys} = 18 \text{ N}$$

Substitute and solve: Combine the above equations and substitute values.

$$18 = (m_1 + m_2)a$$

$$18 = (2 + 4)a$$

$$a = 3 \text{ m/s}^2$$

(B) Determine the magnitude of the force between the masses.

WHAT'S THE TRICK?

According to Newton's third law, the two interacting masses push on each other with opposite and equal force. Therefore, solve for the force on either mass. The forces acting between the blocks are normal forces since the surfaces of the masses push against each other.

Orient the problem: Again, the motion is horizontal, and only horizontal force vectors contribute forces for this motion.

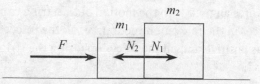

Determine the type of motion: Acceleration obeys Newton's second law.

$$\Sigma F_1 = m_1 a \qquad \text{and} \qquad \Sigma F_2 = m_2 a$$

Work with the masses individually and not as part of a system.

Sum the force vectors in the relevant direction: Force F and the normal force of m_2, which is N_2, push on m_1. Only the normal force of m_1, which is N_1, pushes on m_2. With only one force acting on it, m_2 will be easier to solve. However, the solution for both masses is shown to prove that either mass can be used.

m_1 m_2

$\Sigma F_1 = F - N_2$ $\Sigma F_2 = N_1$

Substitute and solve: Substitute $m_1 a$ for ΣF_1 and $m_2 a$ for ΣF_2. Then, substitute values and solve. The acceleration, $a = 3 \text{ m/s}^2$, was determined in part (A).

$$m_1 a = F - N_2$$
$$(2 \text{ kg})(3 \text{ m/s}^2) = (18 \text{ N}) - N_2$$
$$N_2 = 12 \text{ N}$$

$$m_2 a = N_1$$
$$(4 \text{ kg})(3 \text{ m/s}^2) = N_1$$
$$N_1 = 12 \text{ N}$$

The normal forces pressing between the masses are opposite and equal.

Pulley Problems

Pulley problems are compound-body problems where more than one mass is connected together by a string draped over a pulley. Two common pulley problems are shown in the following diagram. Figure 5.10(a) depicts an **Atwood machine** designed by George Atwood to test constant acceleration and Newton's laws of motion. Figure 5.10(b) shows a modified Atwood machine.

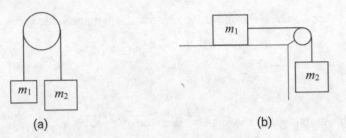

(a)   (b)

FIGURE 5.10 An Atwood machine

Pulleys are used to change the direction of force. The pulleys will be massless and frictionless. The strings will also have zero mass.

Although the masses in pulley problems move in different coordinate directions, they do share one common motion. The masses in pulley problems always move in the same direction as the string. Any force aligned with the motion of the string can influence acceleration. When summing the forces in the relevant direction, set all the forces pointing in the direction of the string's motion as positive and all the forces opposing the string's motion as negative.

<hr>

EXAMPLE 5.5

Atwood Machine

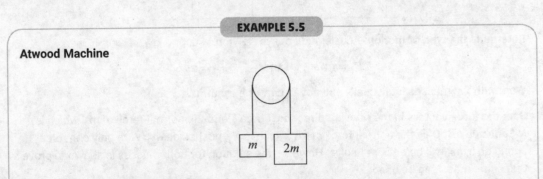

Masses m and $2m$ are connected by a string, which is draped over a pulley. The masses are released from rest. Determine the magnitude of acceleration of mass m.

WHAT'S THE TRICK?

When solving for pulley problems, determine the direction the string is moving. Make this the positive direction and determine all forces parallel to the string. If solving for acceleration, treat both masses as a single system.

Orient the problem: Although the force of gravity acts on both masses, gravity pulls down the $2m$ mass. This causes the smaller mass m to move upward. The diagram below is not a formal free-body diagram. However, viewing the problem in this manner shows a consistent direction of motion that makes it easier to sum the forces for the system.

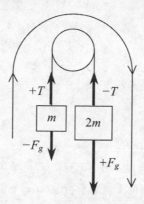

Forces in the direction of motion increase the speed of an object. They have a positive influence on acceleration and are set as positive. Vectors opposing motion slow down an object and are set as negative.

Determine the type of motion: The forces are unbalanced, resulting in the acceleration of the entire system. Applying Newton's second law of motion to the entire system results in the following:

$$\Sigma F_{sys} = (m_1 + m_2)a$$

Sum the force vectors in the relevant direction: Include all the forces parallel to the motion of the string. The sum of the vector forces will be

$$\Sigma F_{sys} = F_{g_m} - T + T - F_{g_{2m}}$$

The pulleys are assumed to be massless. Under these ideal conditions, the tension in a string will be the same everywhere. The equal and opposite tensions will cancel. As a result, tensions can be ignored when solving for the acceleration of the entire system. (Important: tensions cancel only when summing the forces for an entire system. Do not cancel tensions when summing the forces acting on a single independent mass.)

$$\Sigma F_{sys} = F_{g_{2m}} - F_{g_m}$$

Substitute and solve: Substitute $(m_1 + m_2)a$ for ΣF_{sys}. Substitute known equations and values. Then solve.

$$\Sigma F_{sys} = F_{g_{2m}} - F_{g_m}$$
$$(m + 2m)a = (2m)g - (m)g$$

In this problem, the numerical values associated with each mass are coefficients rather than subscripts. Mass $2m$ is not a mass with a value of 2 units. Instead, it is a mass that is twice as large as mass m. This problem will not have a numerical answer. Instead, the solution will be a simplified equation using variables.

$$3ma = 2mg - mg$$
$$a = g/3$$

CHAPTER 6

Circular Motion

Learning Objectives

In this chapter, you will learn how to:

○ Describe the characteristics of uniform circular motion

○ Distinguish between the period and frequency of a circling object

○ Determine the magnitude and direction of the tangential velocity and centripetal acceleration for objects in circular motion

Forces acting perpendicular to the velocity of an object will change the direction of the object. In problems involving the currents in rivers, the force of the wind, and projectiles in motion, the perpendicular force always acts in a constant direction. As a result, motion can be split into independent x- and y-components, which can be analyzed as two separate linear motions taking place simultaneously. However, circular motion presents unique challenges. In circular motion, the direction of force is continually changing and always acts perpendicular to the motion of an object. Therefore, circular motion requires a unique set of equations and its own problem-solving strategies.

Table 6.1 lists the variables and units that will be discussed.

TABLE 6.1 Variables Used in Circular Motion

New Variables	Units
T = Period	s (seconds)
f = Frequency	Hz (hertz) = 1/seconds
v_T = Tangential velocity	m/s (meters per second)
a_c = Centripetal acceleration	m/s² (meters per second squared)
F_c = Centripetal force	N (newtons)
ω = Angular velocity	rad/s (radians per second)

Uniform Circular Motion

Uniform circular motion involves objects moving at a constant speed but with changing velocity. When an object moves at constant speed, the magnitude of velocity is also constant. How then is velocity changing in circular motion? Velocity is a vector quantity, meaning it consists of both magnitude and direction. Objects moving in circular motion are continually changing direction. As a result, the velocity of the object is changing even though it moves at a constant speed.

Like velocity, the acceleration of an object in uniform circular motion has constant magnitude but has changing direction. Acceleration is the rate of change in velocity (the change in velocity during a time interval). If velocity has a constant magnitude, the acceleration will also have a constant value. However, the direction of acceleration constantly changes as an object moves in circular motion. In uniform circular motion, the direction of the acceleration vector is toward the center of the circle. This acceleration is said to be uniform: having a constant magnitude and applied in the same manner (toward the center) at all times.

Period and Frequency

All linear-motion quantities are based on the linear meter. However, the linear meter is not very useful when describing motion that does not follow a straight path. All circles have one thing in common. Objects moving in circles return to the same location every time they complete one **cycle** (one circle, one revolution, one rotation, and so on). A cycle consists of one circumference, and this is the basis for all circular-motion quantities.

The time to complete one cycle is known as the **period**, T. The period of a circling object can be calculated by dividing the time of the motion, t, by the number of cycles completed during time t.

$$T = \frac{t}{\text{number of cycles}}$$

The number of cycles does not have any units, and the units of a period are seconds.

The **frequency** of an object is the number of cycles an object completes during one second. Think of the frequency as how frequently the object is cycling. Mathematically, frequency is the inverse of the period. The units of frequency are inverse seconds (1/s or s^{-1}). These units are also known as Hertz (Hz). Any of these units may be used, and the formula for frequency is simply the inverse of the formula for the period.

$$f = \frac{\text{number of cycles}}{t}$$

The relationship between the period and frequency is expressed in the following equation:

$$T = \frac{1}{f}$$

EXAMPLE 6.1

Period and Frequency

An object completes 20 revolutions in 10 seconds. Determine the period and frequency of this motion.

WHAT'S THE TRICK?

A revolution is another way to indicate a cycle, and a cycle is simply an event. The number of cycles is just the count of an event, and therefore it has no units. Period is the time for one complete cycle. Time has the units of seconds, and therefore time must be in the numerator.

$$T = \frac{t}{\text{number of cycles}} = \frac{10 \text{ s}}{20} = 0.5 \text{ s}$$

Frequency is the inverse formula.

$$f = \frac{\text{number of cycles}}{t} = \frac{20}{10 \text{ s}} = 2 \text{ Hz} \qquad \text{or} \qquad f = \frac{1}{T} = \frac{1}{(0.5)} = 2 \text{ Hz}$$

Tangential Velocity and Centripetal Acceleration

Velocity is not constant in uniform circular motion. However, when any moving object is paused (frozen for an instant of time), it will have an instantaneous velocity vector with a specific magnitude and direction. When objects follow a curved path, the instantaneous velocity is tangent to the motion of the object. Thus, the instantaneous velocity is referred to as the **tangential velocity**. Several instantaneous velocity vectors are shown for the object circling in Figure 6.1.

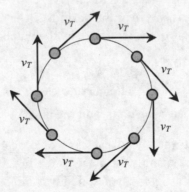

FIGURE 6.1 Instantaneous velocity vectors

In Figure 6.1, it is apparent that although the direction of the tangential velocity is continually changing, its magnitude remains constant. The magnitude of the tangential velocity is also equal to the speed of the circling object. They are both determined using a modified version of the constant-speed formula.

$$v = \frac{d}{t} \quad \text{becomes} \quad v = \frac{2\pi r}{T}$$

As previously stated, circular motion is based on one complete cycle. In one complete cycle, an object travels one circumference, $d = 2\pi r$, in one period, $t = T$.

One important aspect of tangential velocity involves objects leaving the circular path. Forces are responsible for creating circular motion. If the forces causing circular motion stop acting, the object will leave the circular path. When this happens, the tangential velocity becomes the initial velocity for the object's subsequent motion. If no forces act on the object, it will move in a straight line matching the tangential velocity at the time of release, as shown in Figure 6.2(a). However, if another force acts on the object, the object will become subject to the new force. In Figure 6.2(b), the object leaves the circle and is acted upon by gravity, causing projectile motion.

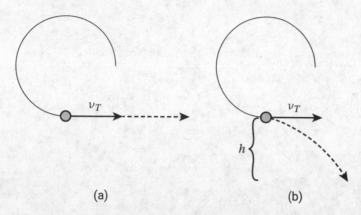

(a)   (b)

FIGURE 6.2 Tangential velocity. In (a), the object is initially circling horizontally on a frictionless surface. In (b), the object is initially circling vertically at a distance h above the surface.

Since circling objects have a changing velocity vector, they are continuously accelerating. This type of acceleration is known as **centripetal acceleration**, a_c. *Centripetal* means "center seeking." Centripetal acceleration is directed toward the center of the circle, as shown in Figure 6.3.

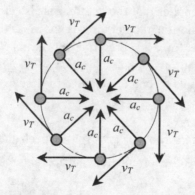

FIGURE 6.3 Centripetal acceleration

Even though circling objects are accelerated toward the center of the circle, their tangential velocities prevent them from ever reaching the center. Since the acceleration vectors lie along the radii of the circle, the centripetal acceleration may also be referred to as the **radial acceleration**. The centripetal acceleration can be determined with the following formula:

$$a_c = \frac{v^2}{r}$$

EXAMPLE 6.2

Tangential Velocity and Centripetal Acceleration

Determine the acceleration of an object experiencing uniform circular motion. It is moving in a circle with a radius of 10 meters and a frequency of 0.25 Hertz.

WHAT'S THE TRICK?

The period is the inverse of the frequency.

$$T = \frac{1}{f}$$

Solving for the tangential velocity requires the period.

$$v = \frac{2\pi r}{T}$$

Solving for the centripetal acceleration requires the tangential velocity.

$$a_c = \frac{v^2}{r}$$

Solve each equation in turn.

$$T = \frac{1}{f} = \frac{1}{0.25\text{Hz}} = 4 \text{ s}$$

$$v = \frac{2\pi r}{T} = \frac{2\pi(10)}{4} = 5\pi \text{ m/s}$$

$$a_c = \frac{v^2}{r} = \frac{(5\pi)^2}{10} = 2.5\pi^2 \text{ m/s}^2$$

Answers may be expressed in terms of π in order to avoid multiplying by 3.14.

Dynamics in Circular Motion

In order for objects to experience acceleration, a net force (sum of forces) must act in the direction of the acceleration. Objects in circular motion are being accelerated toward the center of a circular path. This implies that the net force is also directed toward the center of the circle. In uniform circular motion, the net force is known as the **centripetal force**, F_c. In linear-motion problems, the net force is represented by ΣF and Newton's second law dictates the relationship between the net force and acceleration: $\Sigma F = ma$. In circular-motion problems, F_c replaces ΣF and Newton's second law applied to circular motion becomes

$$F_c = ma_c$$

Frequently, circular-motion force problems involve the speed and/or tangential velocity of the circling object. Since $a_c = \frac{v^2}{r}$, the formula for centripetal force is frequently written as

$$F_c = m\frac{v^2}{r} \qquad \text{or} \qquad \frac{mv^2}{r}$$

Angular Velocity

Angular velocity, ω, is the rate of angular displacement.

$$\omega = \frac{\Delta \theta}{t}$$

The units of angular velocity may appear as revolutions per second (rev/s), revolutions per minute (rpm), or radians per second (rad/s). Working in radians per second has many advantages, and switching to these units involves unit conversion.

$$1\frac{\text{rev}}{\text{s}}\left(\frac{2\pi \text{ rad}}{1 \text{ rev}}\right) = 2\pi \frac{\text{rad}}{\text{s}} \qquad 1\frac{\text{rev}}{\text{min}}\left(\frac{2\pi \text{ rad}}{1 \text{ rev}}\right)\left(\frac{1 \text{ min}}{60 \text{ s}}\right) = \frac{\pi}{30}\frac{\text{rad}}{\text{s}}$$

Angular velocity is also related to the tangential velocity, v, of a point located a radius, r, from the axis about which the object circles.

$$v = \omega r$$

However, this equation requires the angular velocity to be in radians per second.

EXAMPLE 6.3

Angular Velocity

An object in circular motion with a radius of 2.0 meters completes 30 revolutions in 10 seconds.

(A) Determine the object's angular velocity.

WHAT'S THE TRICK?

The object's angular displacement is given as 30 revolutions. Solve the angular velocity.

$$\omega = \frac{\Delta \theta}{t} = \frac{30}{10} = 3 \text{ rev/s} = 3 \text{ s}^{-1}$$

Note: Revolution (rev) is technically a count of a repetitive event and is not actually a unit. It may appear in the units of some problems and be omitted in others. Both are shown above.

(B) Determine the object's tangential velocity.

WHAT'S THE TRICK?

To switch to tangential velocity, the angular velocity must be in radians per second.

$$3.0 \frac{\text{rev}}{\text{s}}\left(\frac{2\pi \text{ rad}}{1 \text{ rev}}\right) = 6.0\pi \text{ rad/s}$$

$$v = \omega r = (6.0\pi \text{ rad/s})(2.0 \text{ m}) = 37.7 \text{ m/s}$$

CHAPTER 7

Energy, Work, and Power

Learning Objectives

In this chapter, you will learn how to:

○ Identify and distinguish among the various forms of energy
○ Understand how constant and variable forces change the energy of an object
○ Calculate and interpret the rate of energy change
○ Analyze energy transformations from one type to another and energy transfers from one object to another

---

Table 7.1 lists the variables that will be discussed.

TABLE 7.1 Variables That Describe Energy, Work, and Power

New Variables	Units
K = Kinetic energy	J (joules) or N · m (newton meter)
U = Potential energy	J (joules) or N · m (newton meter)
E = Total mechanical energy	J (joules) or N · m (newton meter)
W = Work	J (joules) or N · m (newton meter)
P = Power	W (watts) or J/s (joules per second)

Mechanical Energy

Several forms of energy are addressed in a first-year physics course. These include the different types of mechanical energy, electrical energy, thermal energy, heat, light energy, and nuclear energy. At this point, only the different types of mechanical energy will be discussed. Mechanical energy is the sum of the kinetic and potential energy of a system. The system is simply the object, or mass, under investigation.

Kinetic Energy

Kinetic energy, K, is the energy possessed by moving objects. If an object with mass m is moving at a speed v, its kinetic energy is

$$K = \tfrac{1}{2}mv^2$$

Kinetic energy is directly proportional to mass, m, and to the square of velocity, v^2. Kinetic energy is a scalar. Therefore, the direction of an object's velocity is not important. Only the magnitude of velocity (speed) is significant. In addition, kinetic energy is always positive.

Potential Energy

Potential energy, U, is the energy possessed by an object based on its position. To possess useful potential energy, an object must be in a position where it will move when released from rest. When the object is released, it gains kinetic energy at the expense of potential energy. If an object has the potential to create kinetic energy, the object has potential energy. There are two mechanical forms of potential energy commonly encountered in physics: gravitational potential energy and elastic potential energy.

Gravitational Potential Energy

Gravitational potential energy, U_g, is due to the position of an object in a gravity field. A mass, m, in Earth's gravity field, g, is pulled toward Earth by the force of gravity, F_g. When the mass is raised to a height of h, the mass is in a position where it will accelerate toward Earth if released. At a height of h, the mass has the potential to create kinetic energy. If the mass is moved to a greater height, the mass will have a greater potential to increase both the final speed and the final kinetic energy. Height is the key factor determining the magnitude of gravitational potential energy.

$$U_g = mgh$$

Gravitational potential energy is a scalar. Only the height is important. Technically, both height and gravitational potential energy can have negative values. This depends on the location defined as zero height. To avoid working with negative heights, set the zero point for height measurements at the lowest point that the object reaches during a problem.

Elastic Potential Energy

Elastic potential energy, U_s, is the potential energy associated with elastic devices, such as springs. If a spring is stretched or compressed a distance of x, a restoring force, F_s, is generated in the spring according to Hooke's law.

$$F_s = |kx|$$

When a spring is displaced a distance of x and is subsequently released, the restoring force accelerates the spring back to its original length. Thus, displacing a spring has the potential to create kinetic energy.

$$U_s = \tfrac{1}{2}kx^2$$

Elastic potential energy is directly proportional to the spring constant, k, and to the square of displacement, x^2. Elastic potential energy is a positive scalar. The direction the spring is displaced (stretched or compressed) does not matter.

Total Mechanical Energy

Total mechanical energy, ΣE, is the sum of the kinetic and potential energies in a system. Total mechanical energy is an instantaneous value. Freeze the action in a problem and assess if the object you are interested in has height (h), has speed (v), and/or is attached to a spring displaced a distance (x). Next, sum the energy equations corresponding to the key variables that appear in the problem. This is most often done at the start and/or end of a problem.

$$E = K + U$$

Regardless of its form, energy is expressed in units of joules (N • m). As will soon be demonstrated, mechanical energy can change forms through a process known as work. Energy is known as a state function. Work is a process through which the state of energy may be changed.

EXAMPLE 7.1

Recognizing and Calculating Mechanical Energy

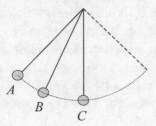

A pendulum initially at rest at point A is released and swings through points B and C. What types of mechanical energy are present at points A, B, and C? What formulas could be used to calculate the total mechanical energy at each point?

WHAT'S THE TRICK?

Look for height (h), speed (v), and spring displacement (x). If any of these are present, include the corresponding energy in the answer.

Point A: The pendulum has height above the lowest point in the swing but it is at rest. Only gravitational potential energy is present. The total mechanical energy at point A is

$$E = U_g = mgh$$

Point B: The pendulum has both speed and height. Both kinetic energy and gravitational potential energy are present. The total mechanical energy at point B is

$$E = K + U_g = \tfrac{1}{2}mv^2 + mgh$$

Point C: The pendulum has lost all its height. Only speed remains, so only kinetic energy is present. The total mechanical energy at point C is

$$E = K = \tfrac{1}{2}mv^2$$

Work

When a force is applied to an object, the force can accelerate the object, changing the object's speed and its kinetic energy. A force can also lift an object to a height h, or it can displace a spring through a distance x. In all these cases, a force displaces an object and changes the object's total energy.

Work is the process of applying a force through a distance to change the energy of an object. Positive work is associated with an increase in the speed of an object. Negative work is associated with a decrease in speed. You can solve for work in two principal ways. The first method involves force and displacement. The second focuses on the change in energy of the object.

Work, W, is the product of the average force, F_{avg}, applied to an object and the component of displacement parallel to the average force, $d_{parallel}$.

$$W = F_{avg}\, d_{parallel}$$

- In most problems, the given force will be equal to the average force needed in the formula. The single exception in introductory physics courses is the restoring force of springs.

- Work requires a change in position known as displacement, which may be indicated in several ways (d, x, or h). It is important to recognize that work involves motion and requires objects to be displaced.

- Both force and displacement are vectors, and vectors at angles can be split into component vectors. The most important aspect of solving for work is remembering to **use parallel force and displacement vectors** or the parallel components of these vectors.

- Forces do work only when they are parallel to the displacement of the object. Forces perpendicular to displacement result in no work.

- Work is positive when the components of force and displacement both point in the same direction. Work is negative when the components of these vectors point in opposite directions.

EXAMPLE 7.2

Solving for Work Using Force and Displacement

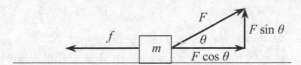

In the figure above, mass m is being pulled horizontally to the right by force F, inclined at angle θ. Force F and its components are shown in the diagram. The surface is rough. Friction, f, opposes the motion.

(A) Determine the work done by the applied force, W_F, as it moves the mass m through displacement d.

WHAT'S THE TRICK?

The applied force, F, is not parallel to the displacement. However, the x-component of the applied force, $F \cos \theta$, is parallel to the displacement and has the same direction as the displacement. $F \cos \theta$ is capable of doing positive work, increasing the speed and kinetic energy of the object.

$$W_F = (+F \cos \boldsymbol{\theta})d$$

(B) Determine the work of friction, W_f.

WHAT'S THE TRICK?

The friction vector, f, is opposite the displacement. It opposes motion, slows the object, decreases the object's kinetic energy, and performs negative work.

$$W_f = (-f)d$$

(C) Determine the net work, W_{net}.

WHAT'S THE TRICK?

The net work is the total work done on the object. One way to solve for the net work is to use the net force (sum of parallel forces, ΣF) acting on the object.

$$W_{net} = (\Sigma F)d$$

$$W_{net} = (F \cos \theta - f)d$$

A second method, using energy, will be discussed in the "Work–Kinetic Energy Theorem" section.

All other forces (force of gravity, normal force, and $F \sin \theta$) are perpendicular to motion. Perpendicular forces and perpendicular components of force do no work.

Work by Gravity

Gravity near the surface of Earth is a good example of the **work done by a constant force**. Earth's gravity, g, is essentially constant for small changes in height. When a force is constant, the average force equals the value of the constant force, $F_{avg} = F_g = mg$. Displacement parallel to the force of gravity is equal to the change in height, $d = \Delta h$. Changes in height are also associated with changes in gravitational potential energy. The work done by gravity can be solved either as a force through a distance or as a change in gravitational potential energy.

$$W = F_{avg}d_{parallel} \qquad\qquad W = -\Delta U_g$$

$$W_g = -mg\,\Delta h \qquad\qquad W_g = -mg\,\Delta h$$

The work done by gravity depends on only a change in height, Δh. As a result, the work done by gravity for horizontal motion is always zero. When objects follow any path consisting of a combination of vertical and horizontal motion, the work done by gravity depends on only the vertical change in height. Although there is a negative sign in the formula, the resulting sign on work also depends on the sign of the change in height, $\Delta h = h_f - h_i$. Ultimately, the work done by gravity is positive for downward motion, when force and vertical displacement have the same direction. It is negative for upward motion, when force and vertical displacement oppose each other.

Most problems about vertical motion involve the **work done by an external applied force that acts to lift an object**. When an object is lifted, the work done by an external applied force, F, is done against gravity. If the object starts at rest and finishes at rest or if the object moves at constant velocity, the work done by the external applied force will have the same magnitude as the work done by gravity. However, work will have the opposite sign.

$$W_F = mg\,\Delta h$$

The work done by an external force to lift an object is positive since force and displacement have the same direction.

EXAMPLE 7.3

The Work of Gravity

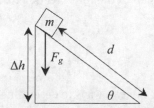

In the figure above, mass m is positioned at the top of a frictionless incline with an angle of θ. The mass slides down the incline a distance d. Determine the work done by gravity.

WHAT'S THE TRICK?

Two displacements are occurring, Δh and d. Use the change in height, Δh, since it is parallel to the force of gravity.

$$W = F_{avg}\,d_{parallel}$$
$$W = F_g\Delta h$$
$$W_g = mg\Delta h$$

The sign on work is positive in this case. Why? You can determine the sign on work using two methods.

- **Vector method:** If the force and the displacement vectors point In the same direction, work is positive. When they are opposite each other, work is negative. F_g and Δh have the same direction. So the work is positive.
- **Energy method:** If the kinetic energy increases during a displacement, the work is positive. Gravity increases the speed of the mass, so the work of gravity is positive.

Work by a Spring

The work by a spring, W_s, is done when a spring is stretched or compressed through a displacement x, thereby changing its length. The work by a spring is a good example of the **work done by a variable force**. Springs obey Hooke's law, $F_s = kx$. The restoring force, F_s, is the instantaneous force in a spring at a specific spring length, x. Changing spring length, Δx, changes the restoring force, as shown in Figure 7.1.

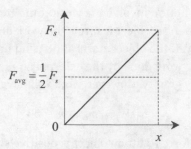

FIGURE 7.1 Restoring force versus displacement

The work formula uses the force acting when distance changes. What force should you use for a spring? Since the restoring force changes in a linear manner, you can use the average force, F_{avg}. The average force is simply half of the force needed to change the length of the spring.

$$F_{avg} = \frac{1}{2}F_s$$

$$F_{avg} = \frac{1}{2}k\,\Delta x$$

As with the work by gravity, there are two ways to determine the work done by a spring. The work done by a spring is the product of the average restoring force, $\frac{1}{2}F_s$, and the change in length of the spring, Δx. It is also equal to the change in elastic potential energy.

$$W = F_{avg}d_{parallel} \qquad\qquad W = -\Delta U$$

$$W_s = -\left(\frac{1}{2}k\Delta x\right)(\Delta x) \qquad\qquad W_s = -\Delta U_s$$

$$W_s = -\Delta\left(\frac{1}{2}kx^2\right) \qquad\qquad W_s = -\Delta\left(\frac{1}{2}kx^2\right)$$

The work done by a spring is positive when the spring is released and moves to restore to its original unstretched or uncompressed length. (In this case, force and displacement have the same direction.) The work is negative when the spring is being stretched or compressed. (In this case, force and displacement oppose each other.)

Problems may instead focus on the **work done on the spring** by an external applied force, F, which acts to stretch or compress the spring from its equilibrium position. The external applied force that stretches or compresses a spring must be equal in magnitude, but opposite in direction, to the resulting restoring force created as the length of the spring changes. As a result, the magnitude of the work done on a spring by an external applied force is equal to the work done by the spring when it is released and moves to restore to equilibrium.

$$W_F = \Delta\left(\frac{1}{2}kx^2\right)$$

The work done on a spring is positive when the spring is being stretched or compressed (force and motion have the same direction). The external applied force must be removed in order for the spring to restore to its original length. Thus, the work done on a spring by an external force is applicable only when the spring is being stretched or compressed.

EXAMPLE 7.4

Work by a Spring

A 4.0-kilogram mass is attached to a spring with a spring constant of 20 newtons per meter. The mass is lowered 0.50 meters to equilibrium, where it remains at rest. How much work was done stretching the spring?

WHAT'S THE TRICK?

This problem involves a spring undergoing a displacement (stretch or compression). Work always involves change. The work done to a spring depends on the change in length of the spring, $\Delta x = x_f - x_i$.

$$W_s = \Delta\left(\frac{1}{2}kx^2\right)$$

$$W_s = \left(\frac{1}{2}kx_f{}^2\right) - \left(\frac{1}{2}kx_i{}^2\right)$$

The spring was not stretched initially, $x_i = 0$.

$$W_s = \left(\frac{1}{2}(20 \text{ N/m})(0.50 \text{ m})^2\right) - (0) = 2.5 \text{ J}$$

The mass attached to the spring is not needed in the formula and is simply a distracter.

Interpreting a Force versus Displacement Graph

Recall that slopes of lines and areas under curves have significance. The slope of a force versus displacement graph (Figure 7.1) has units of N/m. These are the same units as the spring constant (k). The area underneath Figure 7.1 has units of N • m, or joules. These are the same units as for work and energy.

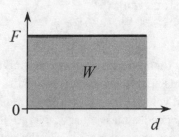

FIGURE 7.2 Force-displacement graph

In the graph in Figure 7.2, the height is equal to the force and the base is equal to the displacement.

$$W = F_{avg} \, d = \text{height} \times \text{base}$$

EXAMPLE 7.5

Work and Force-Displacement Graphs

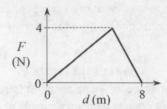

A variable force acts on a 5.0-kilogram mass, displacing the mass 8.0 meters. The force and displacement are graphed above. Determine the work done on the mass by the variable force.

WHAT'S THE TRICK?

Determine the area of the triangle formed by the graph.

$$W = \frac{1}{2}(\text{height} \times \text{base})$$

$$W = \frac{1}{2}(4)(8) = 16 \text{ J}$$

Solving this example required you to find the area of a triangle in a graph. The one-half in the formula actually solves for the average force.

Work–Kinetic Energy Theorem

The **work–kinetic energy theorem** states that work is equal to the change in kinetic energy. When one or more unbalanced forces act on a mass, the mass will accelerate, changing in both speed and kinetic energy. Several forms of work may be present (W_F, W_g, W_s, and W_f) and can be calculated. The net work, W_{net}, done on the mass will be equal to the change in kinetic energy, ΔK, resulting from the total work.

$$W_{net} = \Delta K$$

Using the work–kinetic energy theorem is a quick way to find the net work done on a mass when the initial and final speeds are known. It can also be used to find the final speed of a mass if the net work and initial velocity are known. Since the net work is tied to changes in kinetic energy and changes in speed, a mass must accelerate in order for net work to be non-zero. Thus, when an object moves at constant velocity, the net work is always equal to zero.

EXAMPLE 7.6

Changes in Speed and Net Work (Total Work)

A 2.0-kilogram mass moving at 4.0 meters per second is acted upon by a force, which does 20 joules of work on the mass. Determine the final speed of the object.

WHAT'S THE TRICK?

An initial speed is given, and a final speed is requested. This implies a change in speed. There are two ways to solve for final speed. One involves force and kinematics. The other involves work and energy. The key in this problem is the 20 joules of work. No value for force is given. Try using the work–kinetic energy theorem.

$$W_{net} = \Delta K$$

$$W_{net} = K_f - K_i$$

$$W_{net} = \left(\frac{1}{2}mv_f^2\right) - \left(\frac{1}{2}mv_i^2\right)$$

$$(20\text{ J}) = \left(\frac{1}{2}(2.0\text{ kg})v_f^2\right) - \left(\frac{1}{2}(2.0\text{ kg})(4.0\text{ m/s})^2\right)$$

$$v_f = 6.0\text{ m/s}$$

Power

Power, P, is the amount of work done over a period of time. It is the rate at which work is done. It is also the rate of energy use or energy generation. It can be calculated using the following equation:

$$P = \frac{W}{t} = \frac{\Delta E}{t}$$

Since $W = Fd$, the above formula can be written as

$$P = \frac{Fd}{t}$$

Distance divided by time should be very familiar to you. It is velocity. Therefore, power can also be expressed as

$$P = Fv$$

Power has the units of watts (W). When you look at the top of a lightbulb, it is labeled in watts. A watt is the rate at which the lightbulb uses energy. Another way to express a watt is a joule per second. A 100-watt lightbulb uses 100 joules of energy every second. Power is a scalar quantity. It has magnitude but no direction.

EXAMPLE 7.7

Determining the Power Required to Lift an Object

Determine the power required to lift a 10-newton crate up to a 400-centimeter-high shelf in 2.0 seconds.

WHAT'S THE TRICK?

Power is the rate at which work is done. Solve for work using the methods described in the previous section, and then divide by time. In this case, force and displacement are given. However, displacement is in units of centimeters. You must first convert centimeters into meters.

$$P = \frac{W}{t}$$

$$P = \frac{Fd}{t}$$

$$P = \frac{(10\text{ N})(4.0\text{ m})}{(2.0\text{ s})} = 20\text{ W}$$

Conservation of Energy

Energy is always conserved in an isolated system. The amount of energy present at the start of a problem must remain constant throughout the entire problem. You must consider some important properties of energy to understand conservation of energy fully.

1. Energy can transform from one type to another.
2. Energy can transfer from one object to another.
3. Energy can leave and enter a closed system as work.

Problems about the conservation of energy will ask you to account for all of the joules of energy being transferred and transformed. The unit for energy, joules, can be used for problems involving kinetic energy, gravitational potential energy, elastic potential energy, electrical energy, light energy, nuclear energy, heat, and work.

The actual values of gravitational potential energy and kinetic energy depend on a comparison to a zero-reference height and a zero-reference velocity in order to quantify an object's height and speed. For objects moving on Earth, the planet itself is a commonly used zero-reference point. Conservation-of-energy problems involving an object, such as a roller coaster, should actually specify a roller coaster–Earth system as opposed to mentioning only the roller coaster by itself. However, in beginning physics problems, Earth as a commonly used zero-reference point is often taken for granted. It is wise to understand that the roller coaster does not have quantifiable mechanical energy without regarding it as part of a larger system.

Conservative Forces

The force of gravity, F_g, and the force of springs, F_s, are examples of conservative forces. When work is done only "on" or "by" conservative forces, and all other forces do no work, then the total mechanical energy of a system remains constant. Take, for example, a roller coaster with respect to Earth. At the top of a hill, the roller coaster will have maximum gravitational potential energy, $U_g = mgh$. Should the roller coaster be moving as well, it will also possess kinetic energy, $K = \frac{1}{2}mv^2$. The sum of these energies, $K + U$, is the total mechanical energy of the roller coaster. As the roller coaster descends the hill, it will lose height and gravitational potential energy. However, the roller coaster will simultaneously gain speed and kinetic energy. During the descent, energy transforms from gravitational potential energy into kinetic energy. The force of gravity is a conservative force. When work is done solely by conservative forces, the total mechanical energy of a system (roller coaster with respect to Earth) remains constant.

$$K_i + U_i = K_f + U_f$$

Nonconservative Forces

When nonconservative forces act, energy transfers into or out of a system. As a result, the total mechanical energy of the system is not conserved. At first glance, this seems to violate the conservation of energy. When energy is gained by the system, the energy comes from the environment. When energy is lost by the system, the energy moves to the environment.

Energy is conserved when the system and environment are examined together. However, most problems deal with only a specific system. When nonconservative forces act, the energy of a system is not conserved. The transfer of energy into and out of a system is known as the work of nonconservative forces. It is equal to the change in energy of the system.

$$W_{\text{nonconservative force}} = \Delta E_{\text{sys}}$$

$$W_{\text{nonconservative force}} = E_f - E_i$$

The most commonly encountered nonconservative force is kinetic friction, f_k. Kinetic friction acts on moving objects, such as the roller coaster described in the section "Conservative Forces." When kinetic friction acts, some of the total mechanical energy of the system (roller coaster) is lost (not conserved). Kinetic friction transfers a portion of the initial energy to the environment as heat, resulting in a lower final energy for the system. As a result, the final kinetic energy and speed are less than they would have been in a frictionless environment. The decrease in kinetic energy when kinetic friction slows a moving object is referred to as **kinetic energy lost**. As with other forms of work, the work done by kinetic friction is equal to the product of the force of kinetic friction and displacement. In addition, the work done by friction is equal to the kinetic energy lost by the system and the resulting heat transfer to the environment.

$$W_f = -f_k d$$

$$|W_f| = K_{\text{lost by the system}} = \text{Heat}_{\text{gained by environment}}$$

CHAPTER 8

Momentum and Impulse

Learning Objectives

In this chapter, you will learn how to:

○ Define momentum

○ Define impulse and examine impulse-momentum theory

○ Solve conservation-of-momentum problems for collisions

○ Determine energy changes during collisions

Table 8.1 lists the variables that will be discussed.

TABLE 8.1 Variables for Momentum and Impulse

New Variables	Units
$\vec{p}$ = Momentum	kg · m/s (kilograms · meters per second) or N · s (newton · seconds)
$\vec{J}$ = Impulse	kg · m/s (kilograms · meters per second), or N · s (newton · seconds)

Momentum

There are two types of momentum—linear momentum and angular momentum.

Linear Momentum

Linear momentum is the product of the mass and velocity of an object.

$$\vec{p} = m\vec{v}$$

Momentum has units of kilograms • meters per second. Momentum is related to the inertia of a moving object. The greater the momentum of a moving object, the more difficult the object is to stop. Momentum is a vector quantity. The direction of the momentum vector is the same as the direction of the velocity of the object.

Total Momentum

Many problems involve a system (more than one object) whose objects move simultaneously. The total momentum of a system can be found by adding the individual momentums of all the objects making up the system.

$$\Sigma \vec{p} = \vec{p}_1 + \vec{p}_2 + \cdots$$

$$\Sigma \vec{p} = m\vec{v}_1 + m\vec{v}_2 + \cdots$$

Note that momentum and velocity are both vector quantities. Vector direction for one-dimensional motion can be annotated with a positive or a negative sign as shown in the formula below.

$$\Sigma p = m_1(\pm v_1) + m_2(\pm v_2) + \cdots$$

By adding plus and minus signs, the vector velocity, $\vec{v}$, becomes a scalarlike quantity, v, allowing simple addition. You must note the direction of velocity and add the correct sign when solving a problem for momentum.

Impulse

Impulse is a force that is applied to an object over a period of time. A kick or a shove would be considered an impulse.

$$\vec{J} = \vec{F}\Delta t$$

From this equation, the units of impulse will be newtons • seconds (N • s). This is also equivalent to the units of momentum, kg • m/s. Both momentum and impulse can be expressed in either of these units. Impulse is a vector quantity, and the impulse vector points in the direction of the force acting on the object.

When analyzing the equation for impulse, it is apparent that if the duration of a collision can be lengthened, the force of the impact can be lessened. For example, air bags in an automobile are designed to increase the amount of time needed for a passenger to come to a complete stop, thereby decreasing the force exerted on the passenger during a collision.

Impulse-Momentum Theorem

When an impulse acts on an object, the momentum of the object will change. Note that impulse does not equal momentum even though their units are the same. Impulse causes and is equal to the change in the momentum of an object. Impulse is similar to work in that they both create a change. Work changes the energy of a mass, and impulse changes the momentum of a mass. Impulse and work are both processes of change, while momentum and energy are both state functions.

$$\vec{J} = \Delta\vec{p} = \vec{p}_f - \vec{p}_i = m\vec{v}_f - m\vec{v}_i = \vec{F}\Delta t$$

From the impulse-momentum theorem, the equation describing Newton's second law of motion can be derived.

$$m\Delta v = F\Delta t$$

$$F = m\frac{\Delta v}{\Delta t} = ma$$

The impulse-momentum theorem and Newton's second law of motion are strongly linked and provide two ways to examine force and motion problems.

EXAMPLE 8.1

Change in Momentum

(A) Determine the change in momentum (impulse) for a 0.5-kilogram lump of clay striking a wall at 15 meters per second.

WHAT'S THE TRICK?

The lump of clay will stick to the wall and come to a stop.

$$\Delta \vec{p}_{clay} = m\vec{v}_f - m\vec{v}_i$$

$$\Delta \vec{p}_{clay} = (0.5 \text{ kg})(0) - (0.5 \text{ kg})(15 \text{ m/s})$$

$$\Delta \vec{p}_{clay} = -7.5 \text{ kg} \cdot \text{m/s}$$

The negative sign in the answer indicates that the change in momentum is opposite the initial velocity. The minus sign may not appear in the available answer choices because problems of this type are often concerned with just the value.

(B) Determine the change in momentum (impulse) for a 0.5-kilogram rubber ball striking a wall at 15 meters per second and bouncing off the wall in the opposite direction.

WHAT'S THE TRICK?

The rubber ball will bounce off of the wall. Unless told otherwise, assume that the final speed of the ball leaving the wall is the same as the initial speed of the ball.

$$\Delta \vec{p}_{ball} = m\vec{v}_f - m\vec{v}_i$$

$$\Delta \vec{p}_{ball} = (0.5 \text{ kg})(-15 \text{ m/s}) - (0.5 \text{ kg})(15 \text{ m/s})$$

Note the minus sign on the final velocity of the ball. Set the initial velocity as positive. Because the ball reversed direction during the bounce, the final velocity is in the opposite direction and must have the opposite sign.

$$\Delta \vec{p}_{ball} = -15 \text{ kg} \cdot \text{m/s}$$

The change in momentum (impulse) for an object that bounces with no loss in speed is twice as large as the change in momentum for an object coming to a stop.

Force-Time Graph

Impulse is equal to the area under a force versus time function. The graph in Figure 8.1(a) is an example of the impulse delivered to a soccer ball when the ball is kicked. However, this graph will probably appear in the simplified form of Figure 8.1(b) to allow you to calculate the area under the function with ease.

In the graphs in Figure 8.1, the force is continually changing. Therefore, the formula for impulse requires you to find the average force, F_{avg}, delivered during the time interval, Δt. The magnitude of impulse in the graph in Figure 8.1(b) is

$$J = \frac{1}{2}(\text{height} \times \text{base})$$

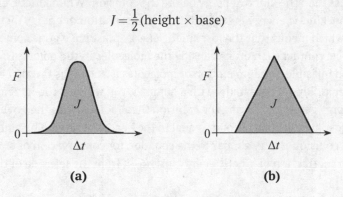

FIGURE 8.1 Force-time graphs

You can solve for impulse in a variety of ways. Impulse is equal to the following:

- The applied force multiplied by the elapsed time, $\vec{F}\Delta t$
- A change in momentum, $m\vec{v}_f - m\vec{v}_i$
- The area under a force-time graph

$$\vec{J} = \vec{F}\Delta t = m\vec{v}_f - m\vec{v}_i = \text{area}_{F\text{-}t \text{ graph}}$$

The most challenging problems may require you to set one formula for impulse equal to another.

Conservation of Momentum

When objects interact in a closed system, the total momentum of the objects is conserved. This means that the total momentum at the beginning of a problem must equal the total momentum at the end of a problem. This is known as the conservation of momentum. It is most often used to solve problems involving collisions and explosions. Mathematically, conservation of momentum can be expressed as

$$\Sigma\vec{p}_i = \Sigma\vec{p}_f$$

$$\vec{p}_{1i} + \vec{p}_{2i} + \cdots = \vec{p}_{1f} + \vec{p}_{2f} + \cdots$$

$$m_1\vec{v}_{1i} + m_2\vec{v}_{2i} + \cdots = m_1\vec{v}_{1f} + m_2\vec{v}_{2f} + \cdots$$

Energy in Collisions

Kinetic energy shares the same variables as momentum. Changes in momentum can cause changes in kinetic energy. When it comes to conservation of energy, only total energy is conserved. Kinetic energy is only one of many energies composing total energy, and kinetic energy can change during a problem. Therefore, kinetic energy is not always conserved.

Kinetic Energy Lost

During a collision where no external force acts, the kinetic energy after the collision will either be equal to or less than the kinetic energy before the collision. When kinetic energy decreases during a collision, kinetic energy is said to be lost. Conservation of energy dictates that energy cannot be lost, which means that the lost kinetic energy must have gone somewhere. During a collision, objects contact each other, causing the molecules in the objects to vibrate. The microscopic, and invisible, random motion of molecules is known as thermal energy. If you touch a nail after striking it repeatedly with a hammer, you will feel the increase in temperature. The thermal energy generated during the collision then radiates into the environment as heat. Kinetic energy lost in collisions becomes equal to the heat generated. The equation for kinetic energy during a collision is very similar to the equation for conservation of energy. The main difference is the subtraction of kinetic energy lost, K_{lost}, from the left side of the equation.

$$K_{1i} + K_{2i} - K_{lost} = K_{1f} + K_{2f}$$

$$\frac{1}{2}m_1 v_{1_i}^2 + \frac{1}{2}m_2 v_{2_i}^2 - K_{lost} = \frac{1}{2}m_1 v_{1_f}^2 + \frac{1}{2}m_2 v_{2_f}^2$$

Collisions are sorted into two main categories, elastic and inelastic, depending on whether kinetic energy is lost during the collision.

Kinetic energy can increase during a collision if previously stored potential energy is released during the collision. Imagine a compressed spring attached to a cart, and the spring is released during a collision with a second cart. As the spring restores to its equilibrium length, it produces an external force that acts on the two-cart system doing work, which increases kinetic energy.

Elastic Collisions

In elastic collisions, objects bounce off of each other. In the process, kinetic energy is conserved. In order for this to occur, the collision must not create any vibrations in the colliding objects, which would imply that the objects never touch each other. Examples may include particles, such as two colliding protons, whose repulsion would prevent them from hitting one another. A larger example would be two objects with a spring mounted on one of them. During the collision, the spring would temporarily store and then release the kinetic energy that would have been lost during the collision. Any energy loss at the spring may be negligible and can be ignored. Very hard objects, such as billiard balls or steel spheres, will also be nearly elastic with minimal energy loss. If a problem states that a collision is elastic, then kinetic energy is conserved and there is no kinetic energy lost. Linear momentum is always conserved in any type of collision where NO external forces, such as friction, act.

$$m_1 \vec{v}_{1i} + m_2 \vec{v}_{2i} = m_1 \vec{v}_{1f} + m_2 \vec{v}_{2f}$$

Inelastic Collisions

If kinetic energy is lost during a collision, then the collision is inelastic. There are two types of inelastic collisions. In the first, the objects may bounce off of each other. In the second, the objects may stick together. When objects stick together, the collision is said to be perfectly (completely, totally) inelastic. The majority of collisions involve an energy loss, making them inelastic collisions.

Inelastic (Ordinary) Collisions

In ordinary inelastic collisions, the objects bounce off of one another as they do in elastic collisions. Momentum is conserved as before, but now kinetic energy is lost.

$$m_1\vec{v}_{1i} + m_2\vec{v}_{2i} = m_1\vec{v}_{1f} + m_2\vec{v}_{2f}$$

$$\frac{1}{2}m_1v_{1_i}^2 + \frac{1}{2}m_2v_{2_i}^2 - K_{lost} = \frac{1}{2}m_1v_{1_f}^2 + \frac{1}{2}m_2v_{2_f}^2$$

Both elastic and ordinary inelastic collisions involve objects that bounce off of each other. How are these two collisions distinguished from each other? Conservation of linear momentum is the same for both of these collisions. The only aspect that differs is kinetic energy.

Perfectly Inelastic Collisions

In these collisions, the objects fuse to become one larger combined mass. To combine, they touch and vibrate, resulting in lost kinetic energy. The conservation of momentum and kinetic energy lost equations are simplified slightly to account for the combined final mass with a single velocity.

$$m_1\vec{v}_{1i} + m_2\vec{v}_{2i} = (m_1 + m_2)\vec{v}_f$$

$$\frac{1}{2}m_1v_{1_i}^2 + \frac{1}{2}m_2v_{2_i}^2 - K_{lost} = \frac{1}{2}(m_1 + m_2)v_f^2$$

EXAMPLE 8.2

Inelastic Collisions

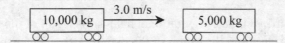

A 10,000-kilogram railroad freight car is moving at 3.0 meters per second when it strikes and couples with a 5,000-kilogram freight car that is initially stationary. What is the resulting speed of the railroad freight cars after the collision?

WHAT'S THE TRICK?

In collisions, momentum is conserved. Since the freight cars combine, this is an inelastic collision.

$$m_1\vec{v}_{1i} + m_2\vec{v}_{2i} = (m_1 + m_2)\vec{v}_f$$

$$(10{,}000 \text{ kg})(3.0 \text{ m/s}) + (1{,}000 \text{ kg})(0 \text{ m/s}) = (10{,}000 \text{ kg} + 5{,}000 \text{ kg})\vec{v}_f$$

$$\vec{v}_f = 2.0 \text{ m/s}$$

CHAPTER 9

Gravity

Learning Objectives

In this chapter, you will learn how to:

○ Understand Newton's law of universal gravitation and its inverse-square-law relationship

○ Visualize the gravity field surrounding masses, such as planets, and calculate its value at specific points in space

○ Calculate circular orbits governed by Newton's law of universal gravitation.

○ Understand Kepler's laws describe orbital motion

On Earth, the gravity at the planet's surface has a constant value of approximately 9.8 meters per second squared (m/s^2) in the downward direction. It can be considered to be a constant field over the surface of Earth with minor fluctuations due to elevation.

The actual value of Earth's surface gravitational field is the result of the mass of Earth and the distance of the surface from the center of the planet. Although Earth has varying surface elevations, these are actually relatively minor. So, it is acceptable to use the constant value of 9.8 meters per second squared for all calculations.

Gravity is an important concept and links together ideas from linear motion as well as circular motion.

Table 9.1 lists the variables that will be discussed.

TABLE 9.1 Variables Used with Gravity

New Variables	Units
F_g = Force of gravity	N (newtons)
G = universal gravitational constant	m^3/kg · s^2 (meters cubed per kilogram seconds squared) or $\frac{\text{N} \cdot \text{m}^2}{\text{kg}^2}$ (newtons · meters squared per kilograms squared)

Universal Gravity

Isaac Newton determined that the **force of gravitational attraction**, F_g, between two masses was directly proportional to the product of their masses and inversely proportional to the square of the distance between them. **Newton's universal law of gravitation** can be expressed as an equation.

$$F_g = G\frac{m_1 m_2}{r^2}$$

The letter G represents the universal gravitational constant, $G = 6.67 \times 10^{-11}$ m^3/kg $\cdot$ s^2.

Inverse-Square Law

As the distance between two masses increases, the force of gravity between them decreases by the square of that distance. This means that a doubling of distance would result in a quartering of the gravitational force between the masses.

EXAMPLE 9.1

Calculating the Change in the Force of Gravity
Calculate the resulting force of gravitational attraction between two masses if one of the masses was to double and the distance between them was to triple.

WHAT'S THE TRICK?
The original force of attraction can be calculated using Newton's law of gravity.

$$F_g = G\frac{m_1 m_2}{r^2}$$

Substituting $2m_1$ for m_1 and $(3r)^2$ for r^2 represents the doubling of the mass and the tripling of the distance, respectively.

$$\left(\frac{2}{3^2}\right)F_g = G\frac{(2m_1)m_2}{(3r)^2}$$

Note that the quantity $3r$ is squared.

$$\left(\frac{2}{9}\right)F_g = \left(\frac{2}{9}\right)G\frac{m_1 m_2}{r^2}$$

The force of gravity between the masses will be two-ninths of its original value.

Gravitational Field

The influence, or alteration, of space surrounding a mass is known as a **gravitational field**. The gravitational field of Earth can be visually represented as several vector arrows pointing toward the surface of Earth, labeled g, as shown in Figure 9.1. The surface gravity field for Earth, g, is 10 meters per second squared. When an object of mass m is placed in Earth's gravity field, it experiences a force of gravity, F_g, in the same direction as the gravity field.

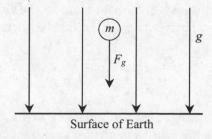

Surface of Earth

FIGURE 9.1 The gravitational field of Earth

The effect on mass m in the gravity field can be solved as a force problem.

$$\Sigma F = F_g$$
$$ma = mg$$
$$a = g$$

It now becomes apparent that the acceleration of a mass in a gravity field equals the magnitude of the gravity field. Although there is a conceptual difference between the gravity field and the acceleration of gravity, they both have the same magnitude and direction.

Finding Surface Gravity

The gravity field of any planet is a function of the mass of the planet and the distance of the planet's surface from its center. The exact relationship can be derived using two formulas for the force of gravity. When a mass m rests on the surface of Earth, the force of gravity can be determined using either the weight formula or Newton's law of gravitation.

$$F_g = mg \qquad \text{or} \qquad F_g = G\frac{mM_{Earth}}{r_{Earth}^2}$$

When these formulas are set equal to one another, mass m cancels. This results in a formula that describes both the strength of the gravity field and the acceleration due to gravity. Note that the mass of Earth is 5.98×10^{24} kg and that the radius is 6.37×10^6 m.

$$mg = G\frac{mM_{Earth}}{r_{Earth}^2}$$
$$g = G\frac{M_{Earth}}{r_{Earth}^2}$$

$$g = (6.67 \times 10^{-11} \text{ N} \cdot \text{m}^2/\text{kg}^2)(5.98 \times 10^{24} \text{ kg})/(6.37 \times 10^6 \text{ m})^2$$
$$g = 9.8 \text{ m/s}^2$$

The gravity field near the surface of Earth is considered uniform. It has the same value at all points close to the surface of Earth. The magnitude of mass m placed on the surface of Earth does not matter since it cancels. Elephants and feathers are both in the same gravitational field and experience the same acceleration of 9.8 m/s^2.

Although the formula for g was derived on the surface of Earth, it can be generalized to solve for gravity on any planet or at a point in space near any planet.

$$g = G\frac{M}{r^2}$$

In its generalized form, M is the mass of the planet and r is the distance measured from the center of the planet to the location where gravity, g, is to be calculated.

Circular Orbits

When a ball is thrown horizontally on Earth, it will follow a parabolic path toward the ground. During its flight, the ball simultaneously experiences a constant downward acceleration and a constant forward velocity. To an observer, it appears that the ball is moving in a parabola relative to a flat Earth.

Newton hypothesized that if a ball could be thrown with sufficient forward velocity, it would travel so quickly that the acceleration pulling it downward would not bring it to Earth. This is because the spherical Earth would curve out of the ball's way as it fell. Today, satellites in orbit are able to do this with speeds exceeding 7,900 meters per second (17,500 miles per hour).

Objects experiencing an acceleration perpendicular to their motion will move in a circle. The magnitude of their acceleration remains constant. However, their direction is constantly changing so that it always points toward the center of rotation.

Calculating Tangential Orbital Velocities

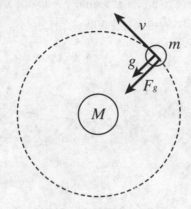

FIGURE 9.2 Satellite of mass *m* orbiting central body *M*

Equations for gravity and circular motion can be combined to determine the velocity of a satellite in a circular orbit. You know following circular-motion equations for centripetal acceleration and centripetal force.

$$a_c = \frac{v^2}{r} \qquad F_c = m\frac{v^2}{r}$$

As seen in Figure 9.2, the acceleration of gravity and the force of gravity are always directed toward the center of the circular path followed by orbiting mass *m*. They are the centripetal acceleration and force on satellites in circular orbits.

$$a_c = g \qquad F_c = F_g$$
$$\frac{v^2}{r} = G\frac{m}{r^2} \qquad m\frac{v^2}{r} = G\frac{mM}{r^2}$$

Rearranging and simplifying two of these equations will solve for orbital speed *v*.

$$v_{\text{orbit}} = \sqrt{\frac{GM}{r}}$$

In the above equation, mass *M* is the mass of a planet or star at the center of an orbiting satellite. Note that the formula does not contain *m*, the mass of the orbiting satellite itself. The speed of a satellite does not depend on its own mass. The satellite's speed depends on the mass of the larger body it is orbiting and on the distance, *r*, from the center of the larger body. There apparently is an inverse relationship between speed and distance. Satellites closer to the central body orbit at higher speeds.

Understanding the formulas associated with orbital motion will help you answer many conceptual problems about changing variables. The key formulas you will need are

$$F_G = G\frac{m_1 m_2}{r^2} \qquad \text{and} \qquad v_{\text{orbit}} = \sqrt{\frac{GM}{r}}$$

Kepler's Laws

Seventeenth-century astronomer Johannes Kepler deduced three laws describing the motion of planets around the Sun.

1. Planets orbit the Sun along an elliptical path where the Sun is at one of the two foci of the ellipse, as shown in Figure 9.3.
2. A line drawn from the Sun to a planet would sweep out an equal area during an equal interval of time. In Figure 9.3, $\text{area}_1 = \text{area}_2$.

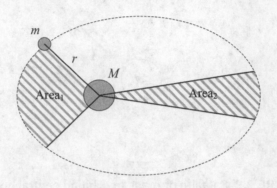

FIGURE 9.3 Kepler's laws

3. The square of the period of a planet's orbit is proportional to the cube of its radius in its orbit about the Sun.

These laws were deduced from Kepler's analysis of observations made by Tycho Brahe, a Danish astronomer and writer. They were considered controversial by several of Kepler's contemporary astronomers. The first law was particularly controversial. Most believed that planets orbited in perfect circles around the Sun.

Kepler's second law illustrates that when a planet is closer to the Sun (known as *perihelion*), it will move at a faster orbital speed than when it is at its farthest point from the Sun (known as *aphelion*). This will allow the imaginary line between Sun and planet to sweep out an equal area during an equal interval of time.

The third law relates a mathematical relationship between orbital period and orbital radius.

$$T^2 \propto r^3$$

Committing this relationship to memory will help you solve conceptual problems about the effect of changing the radius. This relationship works for all objects orbiting a common central mass.

CHAPTER 10

Electric Fields

Learning Objectives

In this chapter, you will learn how to:

- Examine the properties of charge
- Explore electric fields caused by charged plates and point sources
- Draw, analyze, and solve problems for uniform electric fields
- Draw, analyze, and solve problems for electric fields of point charges

Charge is a property primarily associated with electrons and protons. Charged objects create an electric field, which in many ways is similar to the gravity field surrounding masses. While gravity fields create a force on objects with mass, the electric field of one charged object creates a force that acts on other charged objects. Visual representations and the mathematical equations for gravity fields and electric fields are nearly identical. However, while gravity fields cause objects to attract each other, electric fields can cause charged objects to attract or repel one another.

Table 10.1 lists the variables that will be discussed.

TABLE 10.1 Variables Involved with Electric Fields

New Variables	Units
q = charge (small point charge) Q = charge (larger charge, charge on plates)	C (coulombs)
e = Charge of an electron	C (coulombs)
$\vec{E}$ = Electric field	N/C (newtons per coulomb) or V/m (volts per meter)
$\vec{F}_E$ = Electric force	N (newtons)
k = Coulomb's law constant or electrostatic constant	N · m²/C² (newtons · meters squared per coulombs squared)

Charge

Charge is characteristic of electrons and protons. It is associated with a variety of electric properties. The magnitude of charge on an electron and on a proton is the fundamental value of charge, $e = 1.6 \times 10^{-19}$ coulombs, where coulombs (C) is the unit of charge. The charge on a proton is positive, while the charge on an electron is negative. When grouped together, the charge on an equal number of protons and electrons will cancel each other. However, if an object contains more protons than electrons or more electrons than protons, the object will have a net charge, q or Q. The variable q is typically used for a small charge, such as a point charge. The variable Q is typically used for a larger charge, such as the charge on plates. Charge, like matter, is conserved. Although the amount of charge remains constant it can move to another location or to another object. Although masses only attract one another, charges are capable of attracting and repelling each other. Opposite charges attract one another, while like charges repel.

Charged Objects

In addition to electrons and protons, there are other charged objects. If the number of electrons and protons in an atom are not equal, the atom has a net charge and is known as an **ion**. For example, the sodium ion, Na^+, has one less electron than a neutral sodium atom. The oxygen ion, O^{2-}, has two extra electrons compared to a neutral oxygen atom. Everyday objects can also contain excess charge.

Charged objects are usually split into two categories in beginning physics: point charges and charged plates. Point charges are spherical in nature and include electrons, protons, ions, and charged spheres. Regardless of the size of a charged sphere, the entire charge can be assumed to be located at a point in the center of the sphere. This includes hollow spheres and solid spheres. Charged plates consist of two metal plates, which are typically flat, parallel, of equal size, and separated by a distance. The two main types of charged objects create very different electrical effects and employ different equations. The sections that follow will compare and contrast these important charged objects.

Charge is always found in exact quantities. All charges are made up of whole numbers of either electrons or protons. Since the charge on each electron or proton is 1.6×10^{-19} coulombs, 6.25×10^{18} electrons or protons total to 1 coulomb of charge.

A neutral object does not mean the absence of charge. All objects are composed of atoms, which contain electrons and protons. Therefore, all objects contain charge. Neutral objects merely contain the exact same number of electrons and protons, and these opposite charges cancel each other.

Conservation of Charge

Although the individual charge on any one object in a problem may vary, the total charge of all the objects will remain constant. Charges can move from one object to another or can flow through a circuit. However, the total charge of a system (all the objects under examination) at the start of a problem will equal the total charge of the system at the end of a problem. In other words, total charge is conserved.

Charging

Charging an object involves moving extra charges onto or off of the object. How this is accomplished depends on whether the substance to be charged is a **conductor** or an **insulator**. A conductor is a substance that holds its electrons loosely. This allows the electrons to move freely throughout the conductor. The best examples of conductors are metals, which are used as wires in electrical circuits to transport electrons. Insulators are substances that hold their electrons tightly. As a result, insulators prevent the motion of charges. Plastics, which are often used to insulate people from electric shock, are a good example of insulators.

Both conductors and insulators can be charged. When a conductor is charged, the excess charges can move throughout the conductor and readily distribute over the entire outer surface of the conductor. When an insulator is charged, the charges stay in the spot where they have been placed.

All charging methods involve the transfer of electrons. Objects that gain electrons acquire a net negative charge, and objects that lose electrons acquire a net positive charge. **Charging by friction** involves rubbing two nonmetallic objects, such as a wool cloth and a plastic rod, against each other. This method can be used to deposit charge on an insulator. Conductors, such as metals, can be charged by conduction or induction. **Conduction** requires two conductors to touch each other. If one conductor has excess mobile electrons, then some of them will transfer to the other conductor. Charging by **induction** is done without physically touching the object that will be charged.

Electric Fields

Charged objects are surrounded by an electric field, E. The electric field is similar to the gravity field surrounding masses. Like gravity, the electric field is a vector quantity having both magnitude and direction. The electric field of a charged object creates an electric force, F_E, on other charged objects located in the field, just as the gravity field of a mass creates a force on other objects with mass.

Visual representations and the mathematical equations for electric fields and gravity fields are nearly identical. However, charges and their surrounding electric fields vary from mass and gravity fields in some unique ways. Gravity fields always point toward the mass responsible for the field. However, electric fields can point either toward or away from the charge depending on the sign of the charge. While gravity fields cause objects to only attract each other, electric fields can cause charged objects to attract or to repel one another. These differences make electric fields a little more complicated than gravity fields. The direction of the electric field and its effect on positive and negative charges is extremely important.

Uniform Electric Fields

Uniform electric fields exist between two parallel plates containing equal but opposite charges. These fields are considered to be uniform since both the magnitude and the direction of the field are the same at all points between the plates.

Visualizing Uniform Fields

Figure 10.1 compares a uniform electric field and a uniform gravity field.

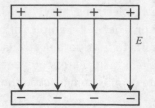 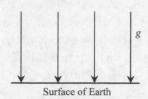

FIGURE 10.1 Electric field of charged plates (left) compared to gravity field (right)

Whereas gravity fields always point toward a mass, such as Earth, electric fields are a bit more complicated because of the existence of two types of charge. An electric field points in the same direction as an electric force points when it is acting on a positive charge. To find the electric field at a point in space due to a charge or to a group of charges, imagine a positive test charge at that location. Determine the direction of force on an imaginary test charge placed in that location. This will be the same as the direction of the electric field. This means that electric fields point away from positive charges and toward negative charges. A uniform field is drawn with parallel and equally spaced arrows. Often, diagrams consist of the arrows only, and the plates responsible for the electric field are not shown.

Magnitude of Uniform Electric Fields

Often, the magnitude of the electric field, E, is given in a problem. Unlike the known value for the gravity field of Earth, $g = 9.8$ m/s^2, the electric field is unique to the plates used in each problem. Although the electric-field strength, E, may be given in a problem, you may also be required to calculate it using the equations that appear in the following sections. The units of the electric field are newtons per coulomb (N/C). The units of the gravity field will most likely be reported in meters per second squared (m/s^2). However, when analyzed as a field rather than as acceleration, the units for the gravity field can be reported as newtons per kilogram (N/kg).

Electric Force in Uniform Electric Fields

If a charge, q, such as a proton or an electron, is placed into a uniform electric field, it will experience an electric force, F_E. This is very similar to placing a mass, m, into the gravity field of Earth. See Figure 10.2.

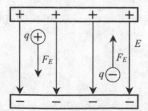 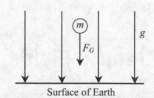

FIGURE 10.2 Force due to uniform fields

The magnitude of force can be determined using the following equations:

Electricity	Gravity
$F_E = qE$	$F_G = mg$

A force is an interaction between an object and an agent. The charged plates are the agent creating an electric field, E. Charge q is the object that experiences a force, F_E. The formula for the magnitude of the electric force is not dependent on whether charge q is positive or negative.

Figure 10.2 clearly shows that the direction of the electric force, F_E, is dependent on the sign of the charge located in the field. The force acting on a positive charge, such as a proton, will be in the direction of the electric field. However, the electric force acting on a negative charge is opposite the field.

Kinematics in Uniform Electric Fields

The sum of all forces acting on an object will determine its resulting motion. The electric force is merely another force. Solve problems involving electric force in the same manner as you solve all other problems involving forces.

1. Orient the problem.
2. Determine the type of motion.
3. Sum the force vectors in the relevant direction.
4. Substitute and solve.

EXAMPLE 10.1

Motion of Charges in Uniform Fields

When an electron is released from rest in a uniform electric field E, it reaches a velocity of v after traveling a distance of x. In terms of v, what will be the electron's velocity if the magnitude of the electric field is doubled while traveling the same distance?

WHAT'S THE TRICK?

This problem combines force and kinematics for a particle that is accelerating. Gravity, while present, is negligible for particles as small as electrons.

Orient the problem: The electric force is the only force included in the calculations. As a result, this problem can be oriented in any manner.

Determine the type of motion: The particle is accelerating. Electrons move opposite the field.

Sum the force vectors in the relevant direction:

$$\Sigma F = F_E$$

Substitute and solve:

$$ma = qE$$

$$a = \frac{qE}{m}$$

The mass and charge of an electron remain constant. Acceleration is directly proportional to the electric field. If the electric field doubles, the acceleration will double.

$$（2a) = \frac{q(2E)}{m}$$

For the speed of an object accelerating from rest, $v_0 = 0$, and moving a set distance, use the following kinematic equation. Solve for v.

$$v^2 = v_0^2 + 2ax$$

$$v = \sqrt{2ax}$$

When distance, x, is held constant, velocity is proportional to the square root of acceleration. Doubling acceleration increases the right side of the equation by the square root of 2. To maintain the equality, velocity must also increase by this factor. The new velocity is $\sqrt{2}\,v$.

$$\sqrt{2}\,v = \sqrt{2(2a)x}$$

Electric Fields of Point Charges

Point charges are charges where the electric field is calculated as though it originates from a single point in space. Obvious point charges are individual electrons, protons, and ions. These tiny objects are essentially points in space. For a larger object to be a point charge, it must be spherical in shape and have the charge evenly distributed over its surface or throughout its volume.

Visualizing Uniform Fields

The field of a point charge radiates outward from the center of the charge and appears similar to the gravity field surrounding a planet viewed as a sphere in space. Figure 10.3 depicts the electric fields of individual positive and negative charges. They are not shown interacting with one another.

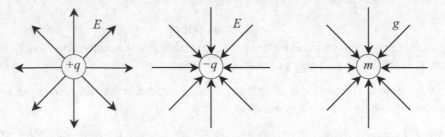

FIGURE 10.3 Electrical fields of positive and negative charges

Although the gravity field of a mass always points toward the mass, the direction of the electric field depends on the sign of charge q. Electric fields point away from positive charges and toward negative charges. The density of the lines indicates the strength of the field. The closer the lines are to each other, the stronger the field is.

When two charges are brought near each other, the electric fields intertwine. Figure 10.4 shows the interactions between unlike and like charges.

Unlike Charges Interacting   Like Charges Interacting

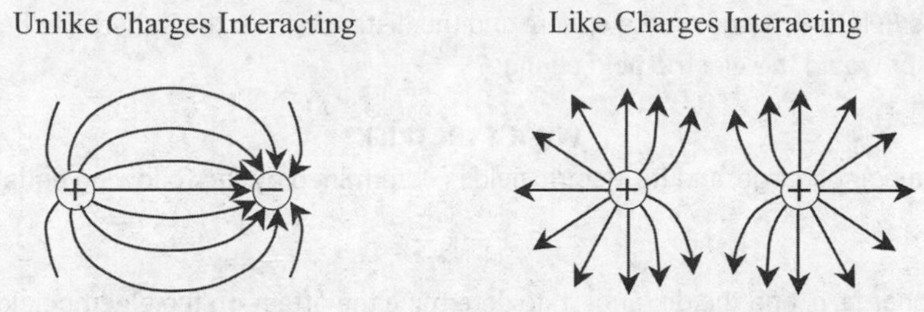

FIGURE 10.4 Interactions between unlike and like charges

Note that had the like charges on the right side of Figure 10.4 been negative, then the field lines would still have followed the same pattern. However, the arrows would have pointed toward the negative charges.

Magnitude of the Electric Field of Point Charges

The magnitude of the electric field, E, of a point charge is solved with an equation that bears a strong resemblance to the equation of the gravity field, g, at a point in space.

Electric Field   Gravity Field

$$E = k\frac{q}{r^2} \qquad\qquad g = G\frac{m}{r^2}$$

The electrostatic constant, $k = 9 \times 10^9$ N • m2/C2, is analogous to the gravity constant, G. Charge q creates the electric field, while mass m creates the gravity field. The relationship between the charge, q, and the electric field, E, is directly proportional. Doubling the size of the charge, q, will double the magnitude of the electric field, E. The distance r is measured from the center of q (or m for gravity) to the point in space where the field is to be calculated. In both of these field formulas, the distance r is inverted and squared. Thus, changes in r are subject to the **inverse-square law**. Doubling the distance r will cause the magnitude of the field to become one-fourth of its original value.

Many problems do not require the entire field to be drawn. Instead, they require a vector direction of the electric field at a specific point in space. This can be accomplished using an **imaginary positive test charge**. Visualize the test charge at the location where field direction is needed. The direction the test charge would move if released is the same as the direction of the electric field. The force on a positive charge is always in the same direction as the field.

EXAMPLE 10.2

Electric Field of Point Charges

A conducting sphere contains a charge q. The electric field at a set distance from the center of the sphere is E. If the charge on the sphere and the distance from the sphere are both doubled, by what factor would the electric field change?

WHAT'S THE TRICK?

A sphere is a point charge, and its electric field is determined by the following equation.

$$E = k\frac{q}{r^2}$$

Double the charge, q, and the distance, r, to determine the effect on the electric field, E.

$$\frac{1}{2}E = k\frac{(2q)}{(2r)^2}$$

The electric field is reduced by a factor of half.

Superposition of Fields and Force

Problems often involve determining the value of the electric field due to more than one point charge. Each charge creates its own electric field that can be calculated using the following formula:

$$E = k\frac{q}{r^2}$$

The formula is solved once for each charge, where r is the distance from the charge being calculated to the point where the field is to be determined. This results in several values for the electric field, one for each charge present. Each calculated electric field is a vector that includes a specific direction. All of the electric-field vectors can be added together using vector mathematics to determine the total electric field.

EXAMPLE 10.3

Electric-Field Superposition

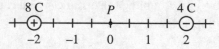

Distance in meters

A +8 coulomb charge is located 2 meters to the left of the origin. A −4 coulomb charge is located 2 meters to the right of the origin. Determine the magnitude and direction of the electric field at point P located at the origin.

WHAT'S THE TRICK?

Identify the +8 C charge as q_1 and the −4 C charge as q_2. Draw and label the electric-field vectors, E_1 for q_1 (away from positive charge) and E_2 for q_2 (toward negative charge) at point P.

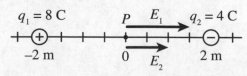

Determine the magnitudes of E_1 and E_2. Magnitudes of vectors are always positive, so the sign on the charges can be ignored when determining electric-field magnitude.

$$|E| = k\frac{q}{r^2}$$

$$|E_1| = k\frac{q_1}{r^2} = (9 \times 10^9 \text{ N} \cdot \text{m}^2/\text{C}^2)\frac{(8 \text{ C})}{(2 \text{ m})^2} = 18 \times 10^9 \text{ N/C}$$

$$|E_2| = k\frac{q_2}{r^2} = (9 \times 10^9 \text{ N} \cdot \text{m}^2/\text{C}^2)\frac{(4 \text{ C})}{(2 \text{ m})^2} = 9 \times 10^9 \text{ N/C}$$

The resultant of adding these vectors is the total electric field due to the superposition of these charges. The direction of these vectors can be seen in the diagram. Since they both point to the right, they can both be regarded as positive vectors.

$$\vec{E} = \vec{E}_1 + \vec{E}_2 = (18 \times 10^9) + (9 \times 10^9) = 27 \times 10^9 \text{ N/C}$$

Electric Force Due to Point Charges

The diagrams in Figure 10.5 show the forces acting on point charges.

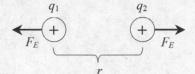
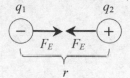

FIGURE 10.5 Forces acting on point charges

Although the interaction between two negative charges is not shown, they would repel each other. The diagram would look very similar to that of the two positive charges shown on the left in Figure 10.5.

When two point charges interact, the resulting force acting on each charge can be determined using **Coulomb's law**. Coulomb's law is extremely similar to Newton's law of gravity.

Coulomb's Law	Newton's Law of Gravity
$F_E = k\dfrac{q_1 q_2}{r^2}$	$F_G = G\dfrac{m_1 m_2}{r^2}$

The magnitude of the electric force, F_E, is determined by multiplying the electrostatic constant, k, by the magnitude of the two interacting charges, q_1 and q_2, and dividing this by the distance between the charges squared, r^2. The magnitude of the electric force is not dependent on the sign of the two charges, and the equation can be solved with all positive values.

The positive and negative signs on charges do influence the direction of the electric force, which can be easily determined by looking at the diagram. Like charges repel, while unlike charges attract. Note that the electric field of each charge is not included in Figure 10.5. Including the field arrows for both charges would have created too much clutter. Always remember that the electric force acting on a positive charge will match the direction of the electric field, while the force acting on a negative charge is opposite the field direction.

Newton's third law of motion is always in effect whenever two objects interact. Two force vectors are shown in each diagram in Figure 10.5. The force on charge 1 (object) is created by the electric field of charge 2 (agent). Similarly, the force on charge 2 (object) is created by the electric field of charge 1 (agent). Coulomb's law solves for the value of both of these force vectors as dictated by Newton's third law: Whenever two objects interact, there is an equal and opposite force between them.

Superposition of Force

When a system consisting of several point charges is present, force vectors add together in a manner similar to the superposition of electric-field vectors described earlier. If three or more charges are present and you must determine the force on one of them due to all the others, use Coulomb's law and superposition. Use Coulomb's law to find the force between the charge in question and every other charge acting on it. In addition, you can find the direction of each of these force vectors using the rules for attraction and repulsion. The result will be several force vectors that can be added together using vector addition to determine the net force acting on the charge in question.

CHAPTER 11

Electric Potential

Learning Objectives

In this chapter, you will learn how to:

○ Explain the nature of electric potential for uniform electric fields

○ Explain the nature of electric potential for point charges

○ Examine the relationship between electric potential and potential energy

○ Understand the motion of charges based on potential

○ Understand capacitors and capacitance

A quantity known as potential, or **electric potential**, is unique to charged objects. Electric potential is associated with electric potential energy, work of the electric force, and conservation of energy. Electric potential, V, is a scalar quantity measured in volts (V). Electric potential, V, can be thought of as an electrical pressure at a point in space. Charges positioned at this point will have electric potential energy and the potential (pressure) to move.

Capacitance is the ability to store charge and electric potential energy. Charged plates are also known as capacitors, and they are useful components in electric circuits. The amount of charge and energy stored by a capacitor depends on its capacitance and on the potential difference between the charged plates.

Table 11.1 lists the variables that will be studied.

TABLE 11.1 Variables Used for Electric Potential

New Variables	Units
V = Electric potential	V (volts)
ΔV = Potential difference	V (volts)
U_E = Electric potential energy	J (joules)
C = Capacitance	F (farads)
ε_0 = Permittivity of free space	8.85×10^{-12} $C^2/N \cdot m^2$
U_C = Energy of a capacitor	J (joules)

Potential of Uniform Fields

Uniform electric fields are a property of equal and oppositely charged parallel plates. Since the electric field between these plates is uniform, the electric potential is evenly distributed in the space between the plates. Figure 11.1 shows two commonly encountered ways to depict the electric potential, V, at various locations between charged plates that have been charged to a 6-volt potential difference, ΔV.

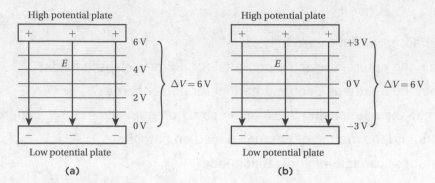

FIGURE 11.1 Electric potential for uniform fields

The magnitude of the electric potential, V, at a location in a uniform electric field can be determined with the following formula:

$$V = Ed$$

In this formula, distance, d, is measured from a location where the potential is equal to zero to the point in the field where potential is to be solved.

Electric potential is analogous to height in mechanics. For charged plates, the value of potential depends on the location of a zero-potential reference line, much as the values of height depend on the location of a zero-height reference line. Commonly, the lowest position an object reaches during its motion is set as zero-height to make mechanics problems easier to solve. Similarly, the same method can be used with uniform electric fields. Potential is a scalar quantity, and the location with the lowest potential can be set as a zero-potential reference line, as seen in Figure 11.1(a). The higher-potential positive plates would then have a potential equal to the value of the potential difference between the plates. However, the zero-reference potential can be set anywhere. Figure 11.1(b) shows the zero-line set midway between the plates. No matter how the values of the potential lines have been quantified, the high and low potential lines must be separated by the value of the potential difference.

However, since the assignment of the zero-potential reference line is arbitrary, it is more important to consider the potential difference, ΔV, between key points when solving problems involving the motion of charges in uniform electric fields. It is the change in potential, ΔV, that will determine the work (change in energy) done on the charges by the electric field. If the electric field, E, and the distance, Δd, that a charge moves through are known, then the potential difference the charge experiences can be determined using the following equation:

$$\Delta V = E\,\Delta d$$

The electric potential can be viewed as lines that are perpendicular to the electric field and equally spaced. These lines are known as **equipotential lines**. Every point on a particular equipotential line has the same equal potential. In Figure 11.2, a point P, a positive charge, and a negative charge are all shown at a location where their electric potentials are 4 volts relative to the negative plate. They are all on the same equipotential line.

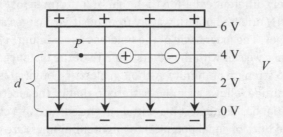

FIGURE 11.2 Electric potential

Some problems will ask for the electric potential of a point or charged object located between charged plates. This is calculated with the same equation used to find the potential of the plates themselves, $V = Ed$. However, the distance, d, is measured from the zero location (usually the negative plate) to the point or charged object, as shown in Figure 11.2.

EXAMPLE 11.1

Electric Potential of a Uniform Field

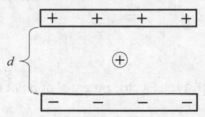

Two charged plates with a 20-newton-per-coulomb electric field are separated by a distance of 10 centimeters. A proton is located at the midpoint between the plates.

(A) Determine the potential difference between the charged plates.

WHAT'S THE TRICK?

To find the potential difference of the charged plates, use the following formula. Remember to change the distance to meters.

$$\Delta V = E\, \Delta d$$

$$\Delta V = (20\ \text{N/C})(0.10\ \text{m}) = 2\ \text{V}$$

(B) Determine the potential acting on the proton.

WHAT'S THE TRICK?

You need to find the potential at the point where the proton is located. Since the proton is at the midpoint between the plates, the distance (in meters) is half the distance between the plates.

$$V = Ed$$

$$V = (20\ \text{N/C})(0.05\ \text{m}) = 1\ \text{V}$$

Potential of Point Charges

The electric potential of point charges is visualized in an entirely different manner and is solved using an entirely different equation than that of uniform fields. The electric field of point charges is radially oriented pointing away from positive charges and radially oriented toward negative charges. The magnitude of the field has a maximum value at the surface of the charge and becomes weaker with the inverse square of the distance from the center of the charge. It has zero strength at infinity. Although electric potential is not a vector and has no direction, its magnitude follows a similar pattern to that of the electric field surrounding a point charge. Potential has its highest magnitude at the surface of a charge and diminishes to zero at infinity. The lines of equal potential (equipotential lines) are perpendicular to the field and form concentric circles around the charge. An example is shown in Figure 11.3 for a positive charge with a 6-volt potential at its surface.

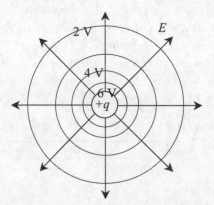

FIGURE 11.3 Electric potential of point charges

The equation for potential is very similar to the equation for the electric field. However, potential is a scalar quantity while the electric field is a vector.

Potential	Electric Field
$V = k\dfrac{q}{r}$	$E = k\dfrac{q}{r^2}$

Positive charges are considered to have a high positive potential, and negative charges have a low negative potential. Although the sign on the charge is important, the distance r from the center of the charge to any point P is set as positive regardless of its direction.

$$V = k\frac{(\pm q)}{|r|}$$

Electric Potential of Several Point Charges

If a problem consists of several point charges surrounding a point P, then merely add together the electric potentials of the individual charges. To find the total potential, simply sum the individual electric potentials.

$$V = V_1 + V_2 + V_3 + \cdots = k\frac{q_1}{r_1} + k\frac{q_2}{r_2} + k\frac{q_3}{r_3} + \cdots$$

Electric Potential Energy

Once the electric potential at a point in space is determined, any charge q can be inserted at that point, and the **electric potential energy**, U_E, can be determined. Simply multiply the electric potential, V, at a point in space by the charge, q, located at that point.

$$U_E = qV$$

The relationship between electric potential and electric potential energy is similar to that between electric field and electric force. A point in space will have a specific electric-field value, E, and a specific electric-potential value, V. When a charge, q, is placed at a point in space, the electric field creates a force, F_E, on the charge. The electric potential can be used to determine its electric potential energy, U_E.

$$F_E = qE \qquad \text{and} \qquad U_E = qV$$

Whether the electric potential is due to charged plates, a single point charge, or several point charges, the equation to find electric potential energy is the same. In Figure 11.4(a), a point charge q is located in the uniform field of charged plates. In Figure 11.4(b), a point charge q_2 is located in the electric field created by point charge q_1.

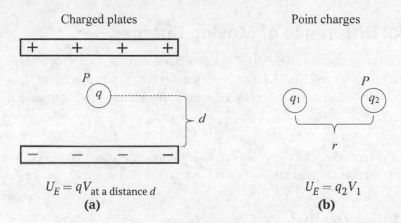

FIGURE 11.4 Electric potential energy

You can substitute the equations for electric potential, V, into the equations found in Figure 11.4(a) and Figure 11.4(b) to create alternate equations for electric potential energy.

$$V = Ed \qquad\qquad V_1 = k\frac{q_1}{r}$$

$$U_E = q(Ed) \qquad\qquad U_E = q_2\left(k\frac{q_1}{r}\right)$$

$$U_E = qEd \qquad\qquad U_E = k\frac{q_1 q_2}{r}$$

The resulting equations solve for electric potential energy without the need to determine electric potential. Electric potential energy is very similar to gravitational potential energy. These derived equations bear strong resemblances to the equations for gravitational potential energy.

$$U_G = mgh \qquad U_G = G\frac{m_1 m_2}{r}$$

Potential energy, U, is the energy of position. When objects are released, they lose their potential energy doing work. The lost energy is usually converted into kinetic energy. As a result, electric potential energy, U_E, can be thought of as the potential to move charges. Electric potential, V, is directly proportional to electric potential energy. In an electrical circuit or in problems involving the motion of charges, electric potential can be thought of as an electrical pressure that exists at a point in space or in a circuit. Any charge located at that point will have the energy needed to move. Electric potential is also commonly called voltage. High voltage is a high potential for charges to move and thus a greater chance of receiving an electric shock.

Motion of Charges and Potential

When charges are released in electric fields, the charges experience a force, causing them to accelerate parallel to electric-field vectors. Positive charges accelerate in the direction of the electric field. Negative charges move opposite the electric field. While in motion, the charges experience a change in potential, known as a potential difference. Changes in speed are associated with changes in kinetic energy, which result from changes in potential energy. Energy is conserved in these processes.

Potential Difference of Moving Charges

When a charge is released in an electric field, the charge moves parallel to electric-field lines, causing a change in distance, Δd (uniform field) or Δr (field of a point charge). The change in distance results in a change in potential, ΔV, specifically known as a potential difference.

$$\Delta V = V_f - V_i$$

In addition, the term *potential difference* can refer to the difference in potential between separated charges, such as the potential between two charged plates.

Work of Electricity

When a charge accelerates, it changes speed and experiences a change in kinetic energy, ΔK. The work–kinetic energy theorem states that changes in kinetic energy are equal to work, $W = \Delta K$. If kinetic energy is changing, another energy in the system must be experiencing an equal but opposite change in order to ensure that energy is conserved. In problems involving a moving charge, the change in kinetic energy is offset by an equal change in electric potential energy. This is directly proportional to the potential difference through which the charge moves.

$$W_E = \Delta K = -\Delta U_E = -q\Delta V = -q(V_f - V_i)$$

Note that the above equation equates the values to one another but does not indicate the correct sign on each value. When a charge speeds up, work and the change in kinetic energy are both positive. In order for kinetic energy to increase, the electric potential energy must decrease. This means the sign on the change in electric potential energy is the opposite of the signs on work and the change in kinetic energy. When a charge slows down, work and the change in kinetic energy are negative while the sign on the change in electric potential energy is positive.

Solving for the correct sign on work using the change in electric potential energy equation, $W = -\Delta U_E = -(q\Delta V)$, is complicated since the equation contains a negative sign and all the variables may be either positive or negative. Most physics problems in beginning courses involve situations where the work of the electric force is positive. As a result, solving for the absolute value of work is actually easier.

$$|W| = |q| \cdot |(V_f - V_i)|$$

EXAMPLE 11.2

Work of Electricity

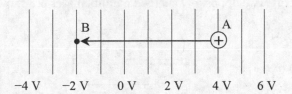

The diagram above shows the equipotential lines in a region of space. Determine the work done on a 2-coulomb charge that is moved from point A to point B.

WHAT'S THE TRICK?

Work is equal to a change in energy. This problem involves equipotential lines. The energy that is changing is electric potential energy, ΔU_E. This is related to the potential difference, ΔV, through which the 2-coulomb charge moves.

$$W = -\Delta U_E = -q\,\Delta V = -q(V_f - V_i)$$

$$W = -\Delta U_E = -q\,\Delta V = -(2\text{ C})[(-2\text{ V}) - (+4\text{ V})]$$

$$W = +12\text{ J}$$

Conservation of Energy

Conservation of energy dictates that the total energy (sum of the kinetic and potential energies) of a closed system must have the same value at any two points during a process.

$$K_1 + U_1 = K_2 + U_2$$

Electricity problems involve electric potential energy, $U_E = qV$.

$$K_1 + U_{E\,1} = K_2 + U_{E\,2}$$

$$\tfrac{1}{2}mv_1^2 + qV_1 = \tfrac{1}{2}mv_2^2 + qV_2$$

Most often, these problems involve an object at rest that accelerates through a potential difference, ΔV, to a point where the object's speed is at the maximum. For plates, maximum speed occurs when a charge moves from one plate with similar charge to the other plate with opposite charge. For a charge moving in the field of a point charge, the maximum speed occurs when the charge reaches infinity. For either of these cases, the maximum kinetic energy equals the change in electric potential energy. The conservation-of-energy formula abbreviates to

$$\tfrac{1}{2}mv_{\text{max}}^2 = q\,\Delta V$$

Capacitors

Plates made of conducting material that are able to store excess charge and electric energy. This property has a useful application for electrical circuits. The capacity of charged plates to hold excess charges is an important factor.

Capacitance

Capacitance, C, which is measured in farads (F), is essentially the ability of a capacitor to store charge and energy. Two oppositely charged parallel plates are shown in Figure 11.5.

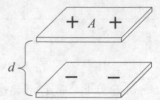

FIGURE 11.5 Oppositely charged parallel plates

The capacitance of the plates is directly proportional to the area, A, of one of the plates and inversely proportional to the distance of the plate separation, d.

$$C = \frac{\varepsilon_0 A}{d}$$

A constant, known as the permittivity of free space, $\varepsilon_0 = 8.85 \times 10^{-12} \text{ C}^2/\text{N} \cdot \text{m}^2$, is needed to turn the proportionality into an equation solving for capacitance, C.

Charging a Capacitor

A battery or a power supply is needed to provide the electric potential needed to charge capacitors or to run circuits. Batteries, like capacitors, are made up of plates of conducting material. The main difference is the presence of chemicals in a battery, which undergo a continuous reaction to keep the plates of the battery loaded with a fixed charge. The fixed charge on the plates of the battery creates a constant potential difference (voltage) across the plates and terminals (ends) of the battery. This potential provides the electrical pressure needed to charge capacitors and/or push charges through circuits. A power supply is a device that functions like a battery. It is plugged into an electrical outlet and adjusts the voltage delivered by the power company to a desired level.

Figure 11.6 shows two versions of the same capacitor connected to a 6-volt battery and uses circuit symbols for the battery and the capacitor.

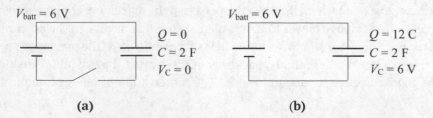

(a)   (b)

FIGURE 11.6 A capacitor connected to a battery

In Figure 11.6(a), the switch is open and the capacitor is initially uncharged, $Q = 0$. An uncharged capacitor has no potential, $V_C = 0$. In Figure 11.6(b), the switch has been closed for a long time, allowing the potential of the battery to push charges onto the capacitor. As the capacitor fills with charge, a potential (pressure) builds up on the plates of the capacitor. The capacitor will continue to fill as long as the potential of the battery is greater than the potential of the capacitor. This process happens quickly at first. As more and more charge builds up on the capacitor, though, forcing additional charges on the capacitor becomes more difficult. Eventually, the potentials become equal ($V_C = V_{batt}$), charging stops, and the capacitor is full. The amount of charge, Q, stored on a capacitor is a function of its capacitance, C, and its potential, V.

$$Q = CV$$

Energy of a Capacitor

Capacitors also store energy. Once a capacitor has been charged, the charges will remain on its plates as long as no pathway is provided for the charges to move from one plate to the other. The battery can even be removed from the circuit and the charges will remain in place on the capacitor. The charges are essentially held in a position to be used later. Therefore, the energy of a capacitor is potential energy, U_c.

$$U_c = \frac{1}{2}QV = \frac{1}{2}CV^2 = \frac{1}{2}\frac{Q^2}{C}$$

EXAMPLE 11.3

Capacitors

A capacitor is connected to a variable power supply with potential V.

Determine the effect on the charge stored on the capacitor if the potential of the power supply is cut in half.

WHAT'S THE TRICK?

The charge on a capacitor is tied to its capacitance, C, and the potential, V, across its plates. When a capacitor is fully charged, its potential will be equal to the potential of the power supply. Capacitance depends on the area, A, of the plates and the distance, d, between them. These quantities are not mentioned in the problem and are therefore assumed to remain constant. When capacitance is constant, the charge, Q, stored on the capacitor is directly proportional to the potential, V, of the capacitor. This potential is being cut in half.

$$Q = CV$$

$$\left(\tfrac{1}{2}Q\right) = C\left(\tfrac{1}{2}V\right)$$

Cutting the potential in half will cut the amount of charge stored in half.

Discharging a Capacitor

A full capacitor stores both charge and energy. If a wire or a circuit is connected between the terminals of the capacitor, the potential difference and stored energy will cause charges to move from one plate of the capacitor to the other. This process occurs very rapidly at first and tapers off as the potential of the capacitor approaches zero. Eventually, the plates of the capacitor will reach a neutral charge.

CHAPTER 12

Circuit Elements and DC Circuits

Learning Objectives

In this chapter, you will learn how to:

○ Define the principal components of a DC circuit

○ Solve problems involving simple DC circuits using Ohm's law

○ Determine heat and power dissipated from a DC circuit

Electric fields induced in a wire will allow current to flow in a circuit. As current flows, it can pass through resistors, illuminate lightbulbs, or accumulate along capacitors. Table 12.1 lists the variables and units that will be discussed.

TABLE 12.1 Variables and Units Used in Circuits

New Variables	Units
R = Resistance	Ω (ohms)
P = Power	W (watts) or joules per second
I = Current	A (ampere) or coulombs per second

Principal Components of a DC Circuit

Battery

The purpose of a **battery** is to create an electric potential difference. Think of a battery as a pump that creates the electrical pressure to push charges through an electric circuit. When a wire is connected between the positive and negative terminals of a battery, charges flow as a current through the wire on their way to the negative terminal of the battery. Potential (voltage) is analogous to height in mechanics. When a mass is released from rest, it moves from a high height to a low height while moving through a distance Δh. If a positive charge is released from rest, it moves from a high potential to a low potential while moving through a potential difference ΔV. Positive charges match the force and energy characteristics of masses in gravity fields. Thus, positive charges are the default charge in electricity. As a result, the convention is to visualize positive charges leaving the positive terminal of the battery (high potential) and flowing through the circuit on the way to the negative terminal (low potential). Although this is how circuits are analyzed, it is not in truth what is really happening. Only electrons are free to move in circuits, and the actual motion of charges is composed of electrons moving in the opposite direction. Mathematically it makes no difference, as protons and electrons have the same magnitude of charge. This only affects direction of charge flow. Even though it is not what is actually happening, the convention is to visualize the flow charges in a circuit as a positive flow.

The potential difference (voltage) between the positive and negative terminals of the battery provides the electrical pressure to the circuit. The overall resistance of the components in the circuit dictates the amount of charge allowed to flow from the battery. The flow of charge is known as the **current**, I, and is measured in units of amperes (A) or coulombs per second.

In circuit diagrams, a long line and a short line represent batteries, as shown in Figure 12.1(a). Sometimes, a battery is represented by a series of long and short lines as seen in Figure 12.1(b). This represents multiple battery cells connected in series, much like a series of batteries lined up end to end in a flashlight.

(a)   (b)

FIGURE 12.1 Batteries

The side of the battery drawn with a long line is the positive terminal. The side drawn with a short line is the negative terminal. When batteries are connected into circuits, positive charges are thought of as departing from the positive terminal, traveling through the circuit, and arriving back at the negative terminal of the battery. Remember: Actual charge flow is electrons in the opposite direction.

Resistors

A **resistor** is a device with a known amount of resistance. As a result, resistors can be used to control current flow in the various branches of a circuit. The **resistance**, R, to current flow has many applications in how a circuit works. As current flow is resisted, some of the electrical energy is dissipated into heat energy in units of joules. Figure 12.2 shows the symbol for a resistor used in circuit diagrams.

FIGURE 12.2 Resistor symbol

Resistors are rated by how much they can resist the flow of current. The actual measurement of resistance is in units of ohms, Ω. Ohm's law can be used to determine the resistance of ohmic resistors.

$$V = IR$$

Using the voltage (V) applied to the circuit and the current (I) flowing through the circuit, you can determine the resistance (R). **Ohmic resistors** are resistors that obey Ohm's law. Resistors that do not obey Ohm's law are said to be nonohmic.

Lightbulbs are essentially resistors that give off light. The principal difference between a plain resistor and a lightbulb is that a lightbulb uses some of the energy dissipated through resistance to produce light. In circuit diagrams, lightbulbs may be represented as resistors. However, the symbol for lightbulbs often varies, as shown in Figure 12.3.

FIGURE 12.3 Lightbulb symbols

Switch

The purpose of a **switch** is to direct current flow around a circuit. In a schematic diagram of a circuit, a switch is represented as a line at an angle to the circuit. Figure 12.4 shows a switch in the open position. When open, a switch blocks the flow of current. When a switch is closed, the circuit becomes complete, and a current is able to flow.

FIGURE 12.4 Symbol for an open switch

DC Circuits

Common problems to solve in physics include circuits with resistors in series or in parallel. Typically, these problems can be categorized as one of three types: series only, parallel only, or series-parallel. Each category has specific problem-solving strategies involved. These strategies will be discussed in this section.

When multiple resistors are used in a circuit, they work together and have a combined resistance known as **equivalent resistance**. Adding resistors results in a single mathematical resistance value that describes a single resistor that is equivalent to the resistors working together. A single resistor possessing this equivalent resistance can replace all the resistors added together. The method of finding equivalent resistance differs for series and parallel circuits.

Series Circuits

In a series circuit, the electrical components are arranged so there is only one path through the circuit. Figure 12.5 shows three resistors arranged in series with a 12-volt battery.

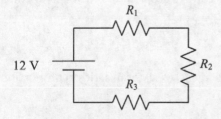

FIGURE 12.5 Series circuit

The equivalent resistance for resistors in series, R_s, is simply the sum of the individual resistances.

$$R_s = R_1 + R_2 + R_3 + \cdots$$

The equivalent resistance for an entire circuit is a single resistance value that represents the total resistance of all the resistors functioning together. This is the resistance that the voltage (electric pressure) produced by the battery must push against. The magnitude of the total current leaving the battery depends on the voltage (electric pressure) produced by the battery and the total equivalent resistance of the circuit, according to Ohm's law.

$$I = \frac{V}{R}$$

As more resistors are added in series, the total resistance increases. Current flow is inversely proportional to resistance. Increasing resistance decreases the total amount of current in a series circuit.

In the complete series circuit shown in Figure 12.5, there is only one pathway. This means the current must flow through every resistor in turn. Think of current as water and the wires and resistors as the pipes carrying the water from the top of a hill (the positive terminal of the battery) to the bottom of a hill (the negative terminal of the battery). Current is composed of charges, and charge is conserved in circuits. The charges leaving the battery as a current must return to the battery. In a series circuit, there is a single path that all the current must follow. With only one pathway available, the current leaving the battery has the same magnitude as the current passing through each of the resistors.

$$I_S = I_1 = I_2 = I_3 = \cdots$$

Although resistors in series all receive the same current, they do not have the same voltage. Voltage (electric potential) is similar to height in Newtonian mechanics. The 12-volt potential difference of the battery in Figure 12.5 can be thought of as a 12-volt hill. The positive terminal of the battery is at the top of a hill, with a value of 12 volts. The negative terminal of the battery is at the bottom of the hill, with a value of 0 volts. As charges flow through the battery, energy from the chemical reaction in the battery increases the voltage of each of the charges from 0 to 12 volts. The action of the battery is analogous to a machine that pulls a roller coaster from zero height to the top of a hill. Voltage is proportional to electric potential energy just as height is proportional to gravitational potential energy. When the charges leave the positive terminal of the battery, they are thought of as flowing downhill toward the negative terminal of the battery, essentially falling through the 12-volt potential difference. On their way to the bottom, the charges pass through circuit components such as resistors. Energy is conserved in circuits, and the resistors in the circuit must use all the energy produced by the battery. As a result, the charges must lose all the voltage given to them by the battery. In a series circuit, the sum of the voltage drops across the resistors must be equal to the voltage increase of the battery.

$$V_S = V_1 + V_2 + V_3 = \cdots$$

In addition, the voltage drop in each resistor in series is proportional to the resistance of the resistors.

EXAMPLE 12.1

Solving Series Circuits

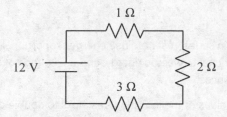

Determine the voltage drop across each resistor.

WHAT'S THE TRICK?

You can use a table like the one below to organize circuit problems. The column headings are the variables for Ohm's law, and any row in the table can be calculated using this law. The components used in the circuit are listed on the left. The top row contains the total values for the entire circuit and is labeled as the battery. The battery produces the total potential for the circuit, pushes against the total resistance in the circuit, and supplies the total current to the circuit. The values shown in bold type are the values given in the original problem. The values in regular type are the values determined during the course of the problem. They are preceded by a number enclosed in parentheses. This is the order in which the problem is solved. Below the table, the four steps are shown in detail.

	V	I	R
Battery	**12 V**	(2) 2 A	(1) 6 Ω
R_1	(4) 2 V	(3) 2 A	**1 Ω**
R_2	(4) 4 V	(3) 2 A	**2 Ω**
R_3	(4) 6 V	(3) 2 A	**3 Ω**

1. Add the resistors to find total equivalent resistance.

$$R_S = R_1 + R_2 + R_3$$

$$R_S = 1\,\Omega + 2\,\Omega + 3\,\Omega = 6\,\Omega$$

2. The battery supplies the total voltage, V_T, that acts as the electric pressure pushing charges against the total equivalent resistance of the entire circuit, R_T. Use Ohm's law to find the total current, I_T, leaving the battery.

$$I_T = \frac{V_T}{R_T}$$

$$I_T = \frac{12\ V}{6\ \Omega} = 2\ A$$

3. The total current leaving the battery must move through each resistor, and is the same in every component in series.

$$I_T = I_1 = I_2 = I_3 = 2\ A$$

4. Use Ohm's law for each resistor.

$$V_1 = I_1 R_1 = (2\ A)(1\ \Omega) = 2\ V$$

$$V_2 = I_2 R_2 = (2\ A)(2\ \Omega) = 4\ V$$

$$V_3 = I_3 R_3 = (2\ A)(3\ \Omega) = 6\ V$$

Compare the ratio of resistance to the ratio of voltage.

$$R_1 : R_2 : R_3 \qquad V_1 : V_2 : V_3$$
$$1 : 2 : 3 \qquad 2 : 4 : 6$$

Energy is conserved, and in series, resistors use energy proportional to their resistance. Voltage is directly proportional to electric energy, and the voltage of resistors in series will be proportional to the resistances.

You can use the voltage values of the individual resistors to double-check your solution. In series, the voltage drops should sum to equal the voltage of the battery.

$$V_T = V_1 + V_2 + V_3$$
$$V_T = 2\,V + 4\,V + 6\,V = 12\,V$$

Parallel Circuits

In a parallel circuit, the electrical components are arranged so there is more than one path through the circuit. Figure 12.6 shows three resistors arranged in parallel with a battery.

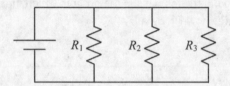

FIGURE 12.6 Parallel circuits

The equivalent resistance for resistors in parallel, R_p, is more complicated than for resistors in series. Parallel resistors are added together so that the sum of their individual reciprocals equals the reciprocal of the total resistance.

$$1/R_p = 1/R_1 + 1/R_2 + 1/R_3 + \cdots$$

Since the above formula solves for the reciprocal of parallel resistance, $1/R_p$, you must remember to invert the answer to determine the equivalent resistance in parallel, R_p. In Figure 12.6, all the resistors are in parallel. The equivalent resistance is also the total resistance that the battery must push against. When resistors are added in parallel, the equivalent resistance is always less than the value of the smallest resistor that was summed. As more resistors are added in parallel, the total resistance decreases, causing the total current in the circuit to increase.

In a parallel circuit, the current moves through multiple pathways. The current flowing through each resistor is a portion of the total current that is produced by the battery. Think of the current as a fixed amount of water leaving the battery. When the water arrives at the junctions, where the paths in the circuit separate, the water will split. Some will flow through resistor 1, some through resistor 2, and the remainder through resistor 3. Then the water will reunite and continue back toward the battery. The total amount of water in the circuit is always the same. When it splits, it must still add up to the total amount that leaves the battery and that later returns to the battery. Therefore, the total current produced by the battery, I_T, is equal to the sum of the current in the parallel resistors. Again, this is based on conservation of charge. Whenever current arrives at a junction (intersection) in the circuit, the current must split in a manner so that the amount of current entering the intersection must equal the amount of current leaving the intersection. The amount of current in the different pathways in a parallel circuit is inversely related to the ratio of resistance. As an example: If there are two parallel pathways and the second path has twice the resistance compared to the first path, then the second pathway will only carry half as much current as the first pathway.

$$I_T = I_1 + I_2 + I_3 + \cdots$$

The potential in a parallel circuit is the same for all resistors. Again, potential (voltage) can be thought of as height, and current can be represented by the flow of water. The battery pushes the water up the hill through a set height. In Figure 12.6, there are three paths, one through each resistor. If water leaves the top of the hill (positive terminal of the battery) and flows through the circuit, it may follow different paths. However, the water must always drop the same amount of height in order to return to the bottom of the hill (negative terminal of the battery). The drop in height must also be equal to the height created by the pumping action of the battery. In a parallel circuit, the voltage drops across each resistor are equal to each other and are equal to the total voltage produced by the battery.

$$V_T = V_1 = V_2 = V_3 = \cdots$$

EXAMPLE 12.2

Solving Parallel Circuits

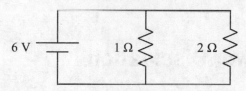

Determine the current in each resistor.

WHAT'S THE TRICK?

As before, use a table to organize the problem.

	V	I	R
Battery	6 V	(2) 9 A	(1) $\frac{2}{3}\,\Omega$
R_1	(3) 6 V	(4) 6 A	1 Ω
R_2	(3) 6 V	(4) 3 A	2 Ω

1. To find the total resistance, add the inverse of each resistor.

$$\frac{1}{R_P} = \frac{1}{R_2} + \frac{1}{R_2}$$

$$\frac{1}{R_P} = \frac{1}{1\,\Omega} + \frac{1}{2\,\Omega} = \frac{3}{2}\,\Omega^{-1}$$

It is very important that you invert this value: $R_T = \frac{2}{3}\,\Omega$

2. Use Ohm's law to find the total current.

$$I_T = \frac{V_T}{R_T}$$

$$I_T = \frac{6\text{ V}}{2/3\ \Omega} = 9\text{ A}$$

3. Voltage remains the same in parallel.

$$V_T = V_1 = V_2 = V_3 = 6\text{ V}$$

4. Use Ohm's law for each resistor.

$$I_1 = \frac{V_1}{R_1} = \frac{6\text{ V}}{1\ \Omega} = 6\text{ A}$$

$$I_2 = \frac{V_2}{R_2} = \frac{6\text{ V}}{2\ \Omega} = 3\text{ A}$$

Compare the ratio of resistance to the ratio of current in each resistor.

$$R_1 : R_2 \qquad I_1 : I_2$$

$$1 : 2 \qquad\quad 6 : 3$$

$$2 : 1$$

When resistance is high, current is low, and vice versa.

Use the current values of the individual resistors to double-check the solution. Charge is conserved. As a result, the current in each parallel pathway should sum to equal the current produced by the battery.

$$I_T = I_1 + I_2 = 6\text{ A} + 3\text{ A} = 9\text{ A}$$

Heat and Power Dissipation

As current flows through resistors, the frictional resistance causes the wires to heat up. The amount of heat, Q, dissipated per second is known as power dissipation. Heat is similar to work in that heat is also a transfer of energy into or out of a system. Heat is the transfer of thermal energy. Be careful when working with heat in electricity problems. The variable Q is used both for heat and for charge. It is possible to mistake them for each other. Power is quantified in units of watts (joules per second).

Joule's Law

Joule's law states that the heat dissipated in a circuit is equal to the current squared multiplied by the resistance and the time that the current flows through the resistor.

$$\text{Heat} = Q = I^2Rt$$

In this case, the variable Q represents heat measured in joules. Joule's law shows that heat is directly proportional to the square of current. If current doubles, heat quadruples.

Power

Power is the rate of change in energy. It is measured in joules per second (J/s), which is known as watts (W).

$$P = \frac{\Delta E}{t}$$

In circuits, the change in energy may involve the electric energy generated (created) by power sources or dissipated (lost) in circuit components, $\Delta UE = QV$. Keep in mind that current is the amount of charge divided by time. Combining these equations creates a useful power equation for use in analyzing electric circuits.

$$P = \frac{\Delta E}{t} = \frac{QV}{t} = IV$$

The resulting equation, $P = IV$, can be combined with Ohm's law, $V = IR$, to create other useful power-equation variations.

$$P = IV = I^2R = \frac{V^2}{R}$$

Most problems can be answered using both $P = IV$ and $V = IR$ as separate equations. However, being aware of all of the power-equation variations does allow for faster problem solving.

CHAPTER 13

Magnetism

Learning Objectives

In this chapter, you will learn how to:

o Examine the properties of magnetism and the types of magnets

o Draw, analyze, and solve problems for uniform magnetic fields

o Draw, analyze, and solve problems for the magnetic fields that surround current-carrying wires

o Analyze how electricity can be generated by changing magnetic flux in order to create electrical pressure, to create emf, and to induce current flow

The source of magnetism and magnetic fields is moving charges. The magnetic field of one magnet (agent) will create a magnetic force on another magnet (object). Like gravity and electricity, magnetic force is a long-range force that acts at a distance even through empty space.

All charges are surrounded by an electric field. When the charges are in motion, they alter the space around them to create the effects known as magnetism. Three objects are surrounded by magnetic fields: fixed (permanent) magnets, current-carrying wires, and individual charges having velocity. Magnetism is also important in electric motors and in the generation of electricity, which is electromagnetic induction.

Table 13.1 lists the variables discussed.

TABLE 13.1 Variables Used in Magnetism

New Variables	Units
$\vec{B}$ = Magnetic field	T (tesla)
$\vec{F}_B$ = Force of magnetism	N (newtons)
μ_0 = permeability of free space	T · m/A (tesla · meters per ampere)
ϕ = magnetic flux	T · m^2 (tesla · meters squared)
ε = emf	V (volts)

Permanent or Fixed Magnets

Fixed magnets are the traditional magnets with which people are familiar. They include bar magnets and horseshoe magnets. Their magnetic properties are the result of electron spin in the orbitals of the magnet's atoms. Although the fixed magnet does not appear to be moving, in fact, the motion of the spinning electrons causes the magnetic field. In most substances, the net spin of all the electrons cancels. In some substances, such as iron, the net effect of the spinning electrons does not cancel. So, each atom acts like a tiny magnet. Groups of atoms having a similar magnetic orientation are known as **domains**. When all the domains of a magnetic substance are aligned, the substance becomes a fixed magnet with a set magnetic field.

The magnetic fields around and between fixed magnets have known properties. The magnetic field surrounding a fixed magnet appears similar to the electric field between an electron and a proton, as shown in Figure 13.1.

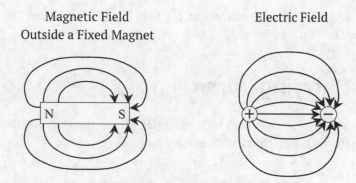

FIGURE 13.1 Magnetic and electric field lines

Fixed magnets have a north pole and a south pole. These poles are analogous to positive and negative aspects of an electric field, but they are not the same thing. Figure 13.1 shows only the magnetic field outside of a fixed magnet, which appears to extend from north to south just as the electric field extends from positive to negative. The electric field terminates on charges. However, the magnetic field forms continuous closed loops. The field lines in the diagram actually extend into and through the fixed magnet. Inside the magnet, the lines run from south to north. However, we are most concerned with the lines outside of fixed magnets, as this portion of the field interacts with other magnets. Fixed magnets can be used to create uniform magnetic fields, as shown in Figure 13.2. The letter $\vec{B}$ represents the vector for magnetic field, which is measured in teslas (T).

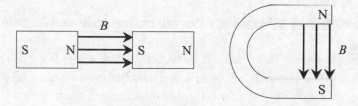

FIGURE 13.2 Uniform magnetic fields

In many problems, the magnets are not drawn. Only the magnetic field is given, or its direction is stated. Quite often, the magnets cannot be drawn as they are not in the plane of the page. The uniform magnetic fields shown in Figure 13.3 are oriented in the z-direction. The magnets that created them are located in front of and behind the plane of this page. Dots are used to indicate a field coming out of the page and ×'s are used to indicate a field going into the page.

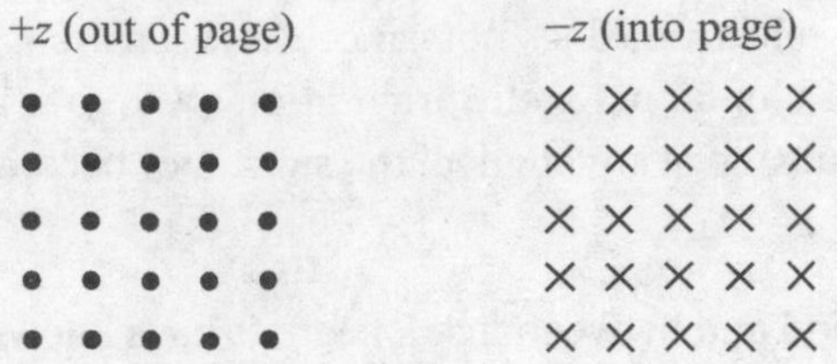

+z (out of page)	−z (into page)
• • • • •	× × × × ×
• • • • •	× × × × ×
• • • • •	× × × × ×
• • • • •	× × × × ×
• • • • •	× × × × ×

FIGURE 13.3 Uniform magnetic fields in the *z*-direction

The even spacing of the symbols representing the field lines indicates that the magnetic fields are uniform. They have the same magnitude and direction at every point.

Current-Carrying Wires

The magnetism of a current-carrying wire is essentially the sum of the magnetism of the moving charges that make up the current in the wire.

Visualizing the Field of Current-Carrying Wires

A wire is essentially a long cylinder. The magnetic field surrounding a current-carrying wire forms concentric circles around every part of the wire. Figure 13.4(a) shows a representation of the circling field. Figure 13.4(b) shows how this field can be rendered two-dimensionally.

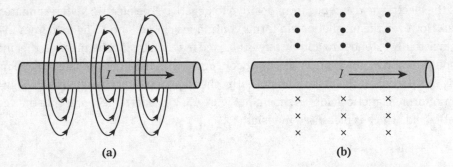

(a)   (b)

FIGURE 13.4 Magnetic fields around a current-carrying wire

Figure 13.4(b) may seem confusing since it shows only a slice of the magnetic field passing through the plane of the page. Above the wire, the field is shown coming out of the page. Below the wire, it is shown entering the page. The directions of the fields above and below the wire are determined by the **right-hand rule**. The thumb of the right hand points in the direction of the current. The curled fingers point in the direction of the circular magnetic field created by the current. When the hand is oriented with the fingers above the wire, the tips of the fingers point out of the page in the $+z$-direction. The field is represented by dots. When the hand is oriented with the fingers below the wire, the tips of the fingers point into the page in the −z-direction, and the field is represented by ×'s. What the field is doing in front of the wire and behind the wire cannot be shown in the plane of the page.

The diagrams in Figure 13.4 can be rotated 90 degrees so that the wire and its current appear to be coming out of or going into the page as shown in Figure 13.5. This diagram clearly shows the circling magnetic fields. Again, the right-hand rule can be used to determine the direction the field is circling. In the left diagram, point the right thumb into the page (×) and the fingers will curl clockwise. In the right diagram, point the right thumb out of the page (dot) and the fingers will curl counterclockwise.

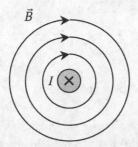 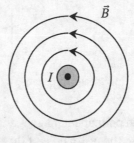

FIGURE 13.5 Using the right-hand rule to determine
the direction of the magnetic field

Magnitude of the Magnetic Field of a Wire

The magnitude of the magnetic field, B, of a long, straight, current-carrying wire is solved with the following equation.

$$B = \frac{\mu_0}{2\pi} \frac{I}{r}$$

The constant, μ_0, is known as the permeability of free space. It has the value $\mu_0 = 4\pi \times 10^{-7}$ T • m/A. The current in the wire is represented by I and is measured in amperes. The distance, r, is measured from the center of the wire to the point where the field is to be solved. The magnetic field is directly proportional to the current in the wire and inversely proportional to the distance from the wire.

EXAMPLE 13.1

Magnetic Field Due to a Current

A long, straight wire carries a current of 2.0 amperes. Determine the magnitude of the magnetic field at a point 10 centimeters from the wire.

WHAT'S THE TRICK?

Use the formula for long, straight wires. Always convert to acceptable units: from centimeters to meters.

$$B = \frac{\mu_0}{2\pi} \frac{I}{r}$$

$$B = \frac{(4\pi \times 10^{-7})}{2\pi} \frac{(2.0)}{(0.10)} = 4 \times 10^{-6} \text{ T}$$

Solenoids and Electromagnets

Coiling conducting wire around a hollow insulating tube creates a device known as a **solenoid**. When current passes through the coils of the solenoid, it generates a magnetic field that resembles the field associated with permanent bar magnets, as shown in Figure 13.6.

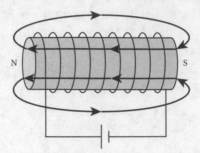

FIGURE 13.6 Magnetic field of a solenoid

Like the field of a permanent magnet, the magnetic field forms loops. Outside the solenoid, the field curves from the north pole of the solenoid toward the south pole. Inside the hollow tube, the field is nearly uniform and extends from the south pole toward the north pole of the solenoid. The solenoid has several advantages over a permanent fixed magnet. It can be turned off, its strength can be adjusted by changing the current passing through it, and the direction of the field can be changed by altering the direction of the current. When an iron core (iron rod) is inserted through the hollow tube, it acts to greatly intensify the strength of the magnetic field, forming a device known as an **electromagnet**.

Force on Moving Charges

Uniform Magnetic Fields and Moving Charges

Tiny moving charges, such as protons and electrons, interact with magnetic fields. These moving charges are so small that the magnetic fields surrounding them are negligible. However, when moving charges move **perpendicularly** to external magnetic fields created by larger magnets, such as fixed magnets and current-carrying wires, the moving charges experience a force of magnetism. The following formula calculates the magnetic force on a charge moving in an external magnetic field.

$$\vec{F}_B = q\vec{v}\,\vec{B}\sin\theta$$

The force of magnetism on a moving charge is the product of its charge (q), its velocity ($\vec{v}$), and the magnetic field ($\vec{B}$), through which it is moving. The formula contains $\sin\theta$, where θ is the angle between the velocity and magnetic-field vectors. The value of $\sin\theta$ is at its maximum and equal to 1 when $\theta = 90°$. Therefore, to receive a maximum force of magnetism, charges must move perpendicularly to the magnetic field. The formula often simplifies to the following.

$$F_B = qvB$$

All three vector quantities, $\vec{F}_B$, $\vec{v}$, and $\vec{B}$, in the above formula must be perpendicular to one another. They each lie on separate axes and require analysis in all the three dimensions (x, y, and z). This means that a charge moving parallel to the field will experience no magnetic forces at all. When charges move parallel to the magnetic field, $\theta = 0°$ and the magnetic force is zero. Zero quantities are often tricky conceptual problems. When charges move parallel to magnetic fields, they are subject to inertia because there is no force of magnetism. If the charges are stationary, they remain stationary. If they are moving at constant velocity, they continue moving at constant velocity.

Electric and gravity fields can cause objects to change speed and/or direction. Magnetic force acting on moving charges is centripetal, resulting in uniform circular motion. Charges moving in magnetic fields are accelerating, but they have constant speed. Figure 13.7 and Table 13.2 compare and contrast the three major uniform fields, the forces they create, and the motion they cause.

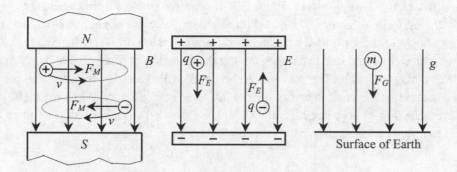

FIGURE 13.7 Three major uniform fields

TABLE 13.2 Forces and Motion Caused by Various Uniform Fields

	Magnetic Field, *B*	Electric Field, *E*	Gravity Field, *g*
Direction	North to south	Positive to negative	Toward Earth
Object	Moving charge, qv	Charge, q	Mass, m
Force	$F_B = qvB$	$F_E = qE$	$F_G = mg$
Resulting motion	Circular motion perpendicular to the field. Positive and negative charges circle in opposite directions as determined by the right-hand rule.	Linear acceleration parallel to the field. Positive charges follow the field, while negative charges move opposite the field.	Linear acceleration in the direction of the field.
Sum of forces	F_C replaces ΣF $$F_C = F_B$$ $$ma_C = qvB$$	$$\Sigma F = F_E$$ $$ma_E = qE$$	$$\Sigma F = F_G$$ $$ma_g = mg$$ $$a_g = g$$
Velocity	Constant speed $$a_C = \frac{v^2}{r}$$	$$v_f^2 = v_i^2 + 2a_E x$$ $$v_f = v_i + a_E t$$	$$v_f^2 = v_i^2 + 2gy$$ $$v_f = v_i + gt$$

The circular motion of moving charges is easier to visualize if the magnetic field is rotated so that it is oriented in the z-direction (out of or into the page), as shown in Figure 13.8.

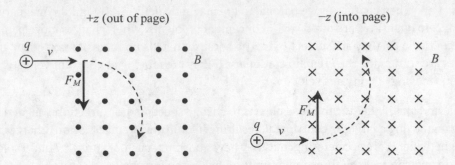

FIGURE 13.8 The magnetic field rotated in the z-direction

Only positive charges are shown in Figure 13.8. If negative charges were shown, they would circle in the opposite direction. The direction, whether clockwise or counterclockwise, that a charge will circle is determined by a slightly different version of the right-hand rule. This version of the right-hand rule is for determining the direction of force on a moving, charged particle by a magnetic field. The previous version of the rule was used to determine the direction of a magnetic field created by the current in a long, straight wire. When magnetic force is involved, the fingers are not curled and should be extended straight, with the thumb and fingers separated by 90° as shown in Figure 13.9.

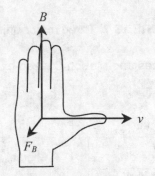

**FIGURE 13.9 Using the right-hand rule
with magnetic force**

To find the direction of force, point the thumb in the direction of the velocity, v, of the moving charge. Point the extended fingers in the direction of the magnetic field lines. The direction that the palm of the hand pushes (a force is a push) is the direction of the force of magnetism. Note that on the left side of Figure 13.9, the palm is pushing out of the page, and the force of magnetism shown is in the positive z-direction.

The right-hand rule gives the direction of force, which can be used to determine whether the object will circle clockwise or counterclockwise. When charges move in a magnetic field, the only force acting on them is the force of magnetism. This force causes uniform circular motion because the magnetic force acts perpendicularly to motion. Therefore, the force of magnetism must point toward the center of the circle. In the left diagram in Figure 13.8, the force of magnetism points downward. Thus, the center of the circle must be below the point where the charge enters the field. The charge must circle clockwise. The opposite is true for the charge entering the $-z$ field in the right diagram in Figure 13.8.

Negative (opposite) charges will always move in a direction opposite that of positive charges. If a positive charge circles clockwise, a negative charge circles counterclockwise. You can handle negative charges in one of two ways.

1. Use the right-hand rule and give the opposite answer.
2. Use the left hand for negative charges.

EXAMPLE 13.2

Charges Moving in Uniform Magnetic Fields

A charge, q, moving at a speed of v enters a uniform magnetic field, B, as shown in the diagram above.

(A) Determine the radius of the circular path in terms of the given variables.

WHAT'S THE TRICK?

In circular motion, centripetal force (F_C) is used instead of ΣF. The force causing the circular motion and pointing toward the center of the circle is the force of magnetism, F_B.

$$F_C = F_B$$
$$m\frac{v^2}{r} = qvB$$
$$r = \frac{mv}{qB}$$

(B) Determine whether the charge shown in the diagram is positive or negative.

WHAT'S THE TRICK?

The center of the clockwise circular motion is below the point where the charge enters the field. The palm of the hand must be oriented downward to push toward the center. Only the left hand is capable of aligning with all three vectors: v, B, and F_B. Therefore, the charge is negative.

Work Done by the Magnetic Force

The force of magnetism does no work on moving charges. The force of magnetism always acts perpendicularly to the motion of charges. In order to do work, a force must be parallel to an object's motion. Since no work is done, the force of magnetism cannot change the kinetic energy and velocity of a moving charge. However, the perpendicular force of magnetism can change the direction of moving charges, causing them to circle at constant speed.

Current-Carrying Wires and Moving Charges

A charge may also be moving through the magnetic field created by a current-carrying wire. The equation for the magnitude of the force of magnetism acting on a charge near a current-carrying wire is the same as when a charge moves in a uniform magnetic field. Although the magnitude of a uniform magnetic field is usually a given, the magnitude of the field surrounding a wire can be determined by using an equation.

$$F_B = qvB \qquad\qquad B = \frac{\mu_0}{2\pi}\frac{I}{r}$$

Determining direction of the force on the charge will require you to use both versions of the right-hand rule. First, you must determine the direction of the magnetic field of the wire. Use the curled-finger version to find the direction of the magnetic field on the side of the wire where the charge is moving. Next, you must determine the force on the moving charge using the straight-finger version and seeing which way the palm is pushing.

Force on Current-Carrying Wires

Any combination of magnets will result in the field of one magnet (agent) creating a force on the second magnet (object). In the preceding section, fixed magnets and currents created forces on moving charges. What about other interactions? The field of a fixed magnet can create a force on a current-carrying wire, causing the wire to move. The field of one current-carrying wire can create a force on another current-carrying wire.

Fixed Magnets and Current-Carrying Wires

If a current-carrying wire passes through an external magnetic field, the field will create a force of magnetism on the wire, causing the wire to move. An external magnetic field is not the field of the object mentioned in the problem. The external magnetic field ($\times$'s) shown in Figure 13.10 is caused by a set of fixed magnets that are positioned in front of and behind the page and cannot be shown in the diagram. This field acts as the agent that puts a force on the wire, the object. The field circling the wire is not shown since the wire does not act on itself.

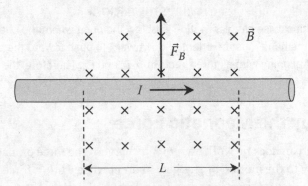

FIGURE 13.10 Current-carrying wire
in an external magnetic field

The force acting on the wire is the product of the current in the wire (I), the length of the wire inside the field ($\vec{L}$), and the external magnetic field ($\vec{B}$). Like the force acting on moving charges, the formula contains $\sin \theta$.

$$\vec{F}_B = I\vec{L}\,\vec{B} \sin \theta$$

As long as the wire is perpendicular to the field, the formula simplifies to the following when solving for vector magnitude.

$$F_B = ILB$$

The direction of this force is determined with the right-hand rule. Currents by definition are considered positive, so the right hand is needed. Use the left hand only if a problem specifically addresses a negative current or the actual electron flow. Since this problem involves force, use the straight-finger version of the right-hand rule. The thumb is the direction of the moving charges, which in this case is the direction of the current, I.

The force of magnetism acting on a current-carrying wire is the basis of all electrical motors. In a motor, the wire is wrapped so that it forms many coils, which extends the length, L, and increases the magnetic force, F_B, of the wire. The coils of wire are arranged so the magnetic force can rotate the coils of wire continuously. The magnetic force in an electric motor converts electrical energy into mechanical energy.

Electromagnetic Induction

The process of generating electricity is known as electromagnetic induction. This is the opposite of the electric motor. In a motor, a current-carrying wire passing through a magnetic field receives a force, causing it to move. In electromagnetic induction, a force is applied, causing a loop of wire or a magnet to move in relation to each other. This induces a current to flow in the wire loop. Induction occurs when the amount of magnetic field, known as **magnetic flux**, ϕ, experiences a change. The change in magnetic flux, $\Delta\phi$, creates an electric potential that induces charges to flow though the loop. When an electric potential is induced, it is known as an **emf**, $\mathcal{E}$, rather than a voltage, and emf has the units of volts. Induction converts mechanical energy into electrical energy.

Magnetic Flux

Magnetic flux, ϕ, can be thought of as the amount of magnetic field, B, passing straight through an area of space, A, as shown in Figure 13.11.

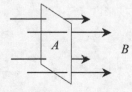

FIGURE 13.11 Magnetic flux

Magnetic flux is the product of the magnetic field and the area through which it passes.

$$\phi = BA$$

In electromagnetic induction, the area used is where the magnetic field and the coil of wire overlap, as shown by the gray areas in Figure 13.12.

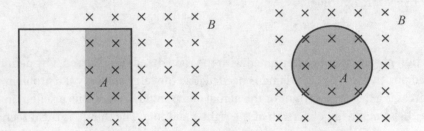

FIGURE 13.12 Electromagnetic induction

The coil of wire can have any shape, but it is often rectangular or circular. It can consist of a single loop of wire or of many loops of wire. When many loops are shown, the wire coil may appear as a spiral helix, where the area, A, is the area of one of the coils.

Inducing a current requires a change in flux, $\Delta\phi$, and this fact is often the basis of conceptual problems. The change in flux can be determined as follows:

$$\Delta\phi = \phi_f - \phi_i$$
$$\Delta\phi = (BA)_f - (BA)_i$$

From the equations, it is apparent that a change in flux results when either the magnetic field or the area change. This change can be accomplished in several ways.

- Moving the magnet and/or the coil toward each other increases the magnetic field, $\Delta\vec{B}$. See Figure 13.13.

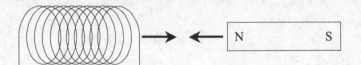

FIGURE 13.13 Coil and magnet moving toward each other

- Moving the magnet and/or the coil away from each other decreases the magnetic field, $\Delta\vec{B}$. See Figure 13.14.

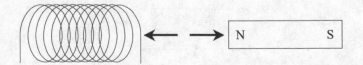

FIGURE 13.14 Coil and magnet moving away from each other

- Spinning a loop of wire in a magnetic field changes the area, ΔA. See Figure 13.15.

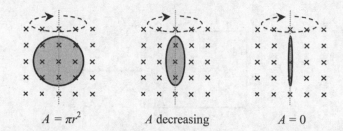

$A = \pi r^2$ A decreasing $A = 0$

FIGURE 13.15 Rotating loop of wire in a magnetic field

- Moving the coil into or out of a magnetic field can change the area, ΔA. See Figure 13.16.

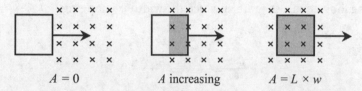

$A = 0$ A increasing $A = L \times w$

FIGURE 13.16 Loop of wire moving into a magnetic field

Emf

Michael Faraday found that if the flux in a loop or coil of wire changes, then a current is created and flows through the loop or coil of wire. This process is known as electromagnetic induction. Changing the flux, $\Delta \phi$, in a closed loop induces (creates) a potential (voltage) in the loop. This special induced voltage is known as an emf and is represented by the variable symbol, ε. It has the units of volts (V). Although it is known by another name, emf acts as any voltage does. The induced emf is the pressure that induces charges to move as a current in the loop of wire. Remember that a change in magnetic flux (a change in magnetic field or area) through a closed loop of conducting material is the necessary event that will induce an emf, ε, which in turn will induce a current to flow in the conducting loop.

The mathematical relationship between changing flux and the induced emf is described by Faraday's law.

$$\varepsilon = \frac{\Delta \phi}{t} \qquad \text{or} \qquad \varepsilon = \frac{\phi_f - \phi_i}{t}$$

The equation of Faraday's law does include a minus sign. However, the minus sign is not needed to determine the magnitude of the induced emf and has been omitted from these equations.

$$\varepsilon = \frac{(BA)_f - (BA)_i}{t}$$

If a rectangular loop of wire moves into or out of a uniform magnetic field at a constant speed, the above formula simplifies greatly. Figure 13.17 depicts a loop moving with a speed of v into a magnetic field of B.

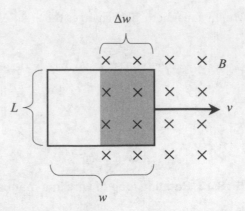

FIGURE 13.17 Loop of wire entering a magnetic field

As the loop moves into the field, the width of the loop exposed to the field is changing, Δw. The speed of the loop equals the rate at which the width is changing.

$$v = \Delta w / t$$

$$\varepsilon = \frac{\Delta \phi}{t} = \frac{B \, \Delta A}{t} = \frac{BL \, \Delta w}{t} = BLv$$

$$\varepsilon = BLv$$

The resulting simplified formula is referred to as the **motional emf** formula. It is commonly used for rectangular loops entering or leaving uniform magnetic fields at constant speed.

Lenz's Law

Generating electricity involves two magnetic fields. The first is the external magnetic field that passes through the loop or coil of wire. The second magnetic field is created when the current begins to flow and turns the wire into a second magnet. Lenz determined that the current in the loop flows in a direction so that the magnetic field of the current opposes the change in the original external magnetic field. There are two possibilities.

1. If the magnetic flux is increasing (magnet and loop moving toward each other or the area of the loop is increasing), the current must flow in a manner so that its magnetic field counters the increase. To counter the increase, the magnetic field of the current in the loop must cancel out the original external magnetic field. In order to accomplish this, the magnetic field of the current must point in the opposite direction of the field created by the external magnet.

2. If the magnetic flux is decreasing (magnet and loop moving away from each other or the area of the loop is decreasing), the current must flow in a manner that increases the magnetic field. In order to accomplish this, the magnetic field of the current in the loop must point in the same direction as the original external magnetic field.

CHAPTER 14

Simple Harmonic Motion

Learning Objectives

In this chapter, you will learn how to:

- Identify the types and terminology of SHM
- Solve problems for springs and pendulums
- Interpret graphical representations of SHM
- Summarize key values in energy and force during SHM

Simple harmonic motion (SHM) is the periodic and repetitive motion of an object, such as a pendulum, a mass on a spring, or a plucked guitar string. In each of these examples of SHM, the object is displaced from its initial point of equilibrium and oscillates in a predictable manner about the equilibrium position.

Terms Related to SHM

Period

Period, T, is the time in seconds for an object in SHM to complete one full oscillation. An example of a full oscillation would be the movement of a pendulum from its initial starting position back to that position again. When more than one oscillation is given, the period can be determined by dividing the total time by the number of oscillations.

$$T = \frac{\text{total time}}{\text{\# of oscillations}}$$

Note that the number of oscillations has no units. It is simply the count of an event.

Frequency

Frequency, f, is the number of full oscillations that occur in 1 second. Divide the number of oscillations by the time interval to determine the frequency.

$$f = \frac{\text{\# of oscillations}}{\text{total time}}$$

Frequency is the reciprocal of period and therefore has units of 1/second, which is also known as Hertz (Hz). The relationship between period and frequency can be expressed as

$$T = \frac{1}{f} \qquad \text{and} \qquad f = \frac{1}{T}$$

Period and frequency are inversely proportional. As one increases, the other decreases.

EXAMPLE 14.1

Determining the Period from the Frequency

A plucked guitar string vibrates at a frequency of 100 Hertz. What is the period of vibration of the string?

WHAT'S THE TRICK?

Period is the reciprocal of frequency.

$$T = \frac{1}{f}$$

$$T = \frac{1}{100 \text{ Hz}}$$

$$T = 0.01 \text{ s}$$

Amplitude

Amplitude, A, is the magnitude of the maximum displacement, $A = x_{max}$, of an oscillating particle or wave relative to its rest position. Amplitude is a measure of the intensity of the oscillation and is directly proportional to the energy imparted into the oscillating system. The amplitude does not affect the period or the frequency.

The amplitude can be assigned as either a positive or negative value, depending on which side of the equilibrium position is being described. For example, the amplitude of a plucked guitar string can either be above or below the equilibrium position of the string at rest. These two maximum amplitudes are individually assigned values of $+A$ and $-A$, respectively.

Oscillations of Springs

Determining the Spring Constant

Imagine a mass suspended by a spring, as represented in Figure 14.1. The force of gravity pulls the mass toward Earth, while the restorative force of the spring pulls the mass upward in an effort to restore the spring to its original, unstretched position. At equilibrium, the mass will be at rest and the spring will be stretched by an amount proportional to the force of gravity upon the mass.

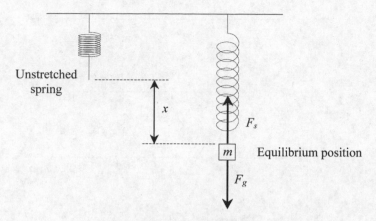

FIGURE 14.1 Hooke's law

The **restorative force** of a spring, F_s, is represented by the following equation, known as **Hooke's law**.

$$F_s = -kx$$

In Hooke's law, x represents the displacement of the spring from its unstretched position, in meters, and k represents the **spring constant**, in newtons per meter. The spring constant, k, is specific to each spring, regardless of the mass suspended from the spring.

The **equilibrium position** is reached when the restorative force of the spring is equal in magnitude but opposite in direction to the gravity force acting on the mass. This can be described by setting the force of gravity equal to but opposite in sign to the restorative force of the spring.

$$F_s = F_g$$

$$kx = mg$$

Determining the Period of a Spring in SHM

The period of a spring in SHM is the amount of time, in seconds, in which the mass on a spring moves from an initial position back to that same initial position. The actual motion of a spring-mass system follows the same linear path up and down or back and forth. However, if the motion of an oscillating spring is graphed versus time, the resulting function is a sine wave. Figure 14.2 is a graph of position versus time. The positions and stretch of a vertically oriented spring-mass system have been superimposed on the graph at five key locations.

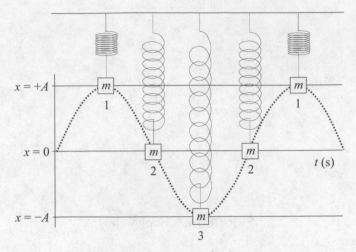

FIGURE 14.2 Position versus time

The sine wave represents the predictable, periodic nature of an oscillating spring-mass system during a time interval. If the mass hanging from a spring were released at position 1 in Figure 14.2, the mass would descend vertically, passing through equilibrium at position 2. The mass would reach maximum displacement at position 3, where it would reverse direction. On the return trip, the mass would pass through equilibrium at position 2 a second time and would return to its starting location at position 1. One complete cycle occurs when the oscillator returns to its initial position (position 1) for the first time. The period is the time of one complete cycle. For a spring-mass system, the period, T_s, can be determined with this formula.

$$T_s = 2\pi \sqrt{\frac{m}{k}}$$

EXAMPLE 14.2

Determining Period and Frequency of an Oscillating Spring

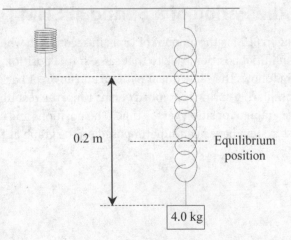

A 4.0-kilogram mass is suspended from an unstretched spring. When released from rest, the mass moves a maximum distance of 0.20 meters before reversing direction. What are the period and frequency of this spring-mass oscillator?

WHAT'S THE TRICK?

For problems involving a spring, you must determine the spring constant first. The mass moves a distance of 0.20 meters before it reverses direction. This is the entire up-down motion. The equilibrium position is in the middle of this motion. So, it occurs at 0.10 m.

$$F_s = F_g$$
$$kx = mg$$
$$k = \frac{mg}{x}$$
$$= \frac{(4.0 \text{ kg})(10 \text{ m/s}^2)}{(0.10 \text{ m})}$$
$$= 400 \text{ N/m}$$

Apply the spring constant to the formula for the period of a spring-mass oscillator.

$$T = 2\pi\sqrt{\frac{m}{k}}$$
$$= 2\pi\sqrt{\frac{(4.0 \text{ kg})}{(400 \text{ N/m})}}$$
$$= 0.20\pi \text{ s}$$

Frequency is simply the reciprocal of period.

$$f = \frac{1}{T}$$
$$= \frac{1}{0.2\pi}$$
$$= \frac{5}{\pi} \text{ Hz}$$

Oscillations of Pendulums

Pendulums are simple harmonic oscillators that take the form of a mass suspended at the end of a string. The period of a pendulum is the time for the mass at the end of the pendulum to oscillate from an initial position back to that initial position again, as shown in Figure 14.3.

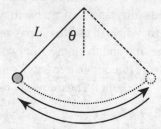

FIGURE 14.3 The period of a pendulum

Pendulums are not perfect simple harmonic oscillators. They approximate simple harmonic oscillators as long as the displacement angle, θ, is small ($\theta \leq 10°$). The period of a pendulum can be determined using the following formula.

$$T_p = 2\pi\sqrt{\frac{L}{g}}$$

As the length of the string increases, the period increases, causing the frequency to decrease. The amount of mass at the end of the string and the displacement from equilibrium (distance the mass is moved sideways) *do not* affect the period of the pendulum at all.

Graphical Representations of SHM

When the displacement of an object in SHM is plotted on a graph of displacement versus time, the result is a sine wave, as shown in Figure 14.4.

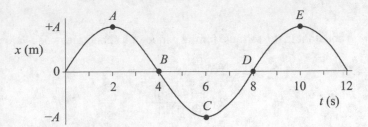

FIGURE 14.4 Displacement versus time

One complete oscillation is represented by the time for the graph to go from point A to point E. The period is the time of one complete cycle.

$$T = t_E - t_A = 10 - 2 = 8 \text{ seconds}$$

The frequency is the inverse of the period and is therefore $\frac{1}{8}$ Hz. This means that one-eighth of an oscillation occurs every second.

Energy in Oscillations

In an oscillation, energy continually changes form from potential energy to kinetic energy. Potential energy depends on displacement. When the oscillator reaches its maximum displacement, the potential energy will also reach its maximum value. At the equilibrium position, the oscillator has a displacement of zero and a potential energy of zero. Kinetic energy is the complete opposite. At maximum displacement, an oscillator has an instantaneous speed of zero and a kinetic energy of zero. When an oscillator passes through the equilibrium point, it attains its greatest speed and has maximum kinetic energy. However, the total mechanical energy, $\Sigma E = U + K$ (potential energy plus kinetic energy), remains constant and is always conserved throughout a complete oscillation. These energy relationships are graphed in Figure 14.5.

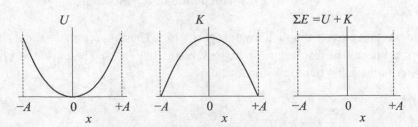

FIGURE 14.5 Energy graphs of oscillations

Conservation of energy is often needed to solve oscillation problems. Since the total energy is constant at every point in an oscillation, the total energy at any point 1 can be set equal to the total energy at any point 2.

$$\Sigma E_1 = E_2$$

$$U_1 + K_1 = U_2 + K_2$$

For a spring, substitute the potential energy of a spring.

$$\frac{1}{2} kx_1^2 + \frac{1}{2} mv_1^2 = \frac{1}{2} kx_2^2 + \frac{1}{2} mv_2^2$$

For a pendulum, substitute the potential energy of gravity.

$$mgh_1 + \frac{1}{2} mv_1^2 = mgh_2 + \frac{1}{2} mv_2^2$$

Force and Acceleration

Acceleration is the result of the sum of the forces acting on an object. Figure 14.6 diagrams the key force and energy relationships for a mass-spring system that is oscillating on a horizontal frictionless surface. In an oscillation, the sum of forces is zero when the object is at the equilibrium position. When the sum of forces is zero, the acceleration is also zero, $\Sigma F = ma$. When the oscillator is moved out of equilibrium (the spring is stretched/compressed or the pendulum is moved to either side), the sum of the forces increases with displacement, reaching its highest value at maximum displacement.

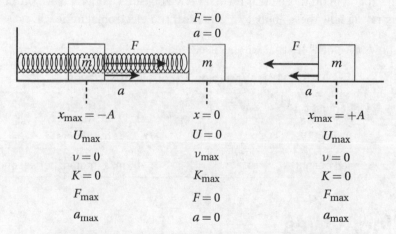

FIGURE 14.6 Force and energy trends in oscillations

CHAPTER 15

Waves

Learning Objectives

In this chapter, you will learn how to:

- Define transverse and longitudinal traveling waves and their behavior in a medium
- Define the properties of mechanical and electromagnetic waves
- Explain the Doppler effect
- Describe superposition and how it leads to standing waves and beats

Oscillations involve a single vibrating object. When energy from one oscillator is transferred to another oscillator, it is known as a wave. The principal function of waves is to transmit energy. Wave behavior can take the form of mechanical waves or electromagnetic (light) waves. Here, we will expand upon those ideas by first reviewing some general wave properties and then demonstrating how these apply to mechanical and electromagnetic waves.

Table 15.1 shows the variables that will be discussed.

TABLE 15.1 Variables That Describe Waves

New Variables	Units
λ = wavelength	m (meters)
c = speed of light	3×10^8 m/s (meters per second)

Traveling Waves

A **wave** is an organized disturbance consisting of many individual oscillators. Although each individual oscillator barely moves, the resulting wave can travel great distances at a constant speed in a specific medium. While traveling, this disturbance transmits energy from one place to another without moving any physical objects over the same distance. The energy propagates outwardly from the source of the oscillations.

There are two types of traveling waves: **transverse** and **longitudinal**. Figure 15.1 illustrates the two types.

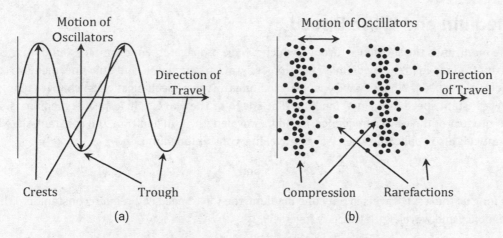

FIGURE 15.1 Transverse (a) and longitudinal (b) waves

In transverse waves, the oscillators vibrate perpendicularly to the direction of wave propagation (they transverse/cross equilibrium). In longitudinal waves, the oscillators vibrate parallel (along or longitudinal) to the direction of wave propagation.

Mathematically, both transverse and longitudinal waves can be graphically represented as sine-wave functions. The longitudinal wave has a unique appearance that is difficult to depict in diagrams. As a result, diagrams used to analyze longitudinal waves depict a sinusoidal form that resembles the appearance of transverse waves. Figure 15.2 identifies the principal components of both transverse and longitudinal waves. The figure depicts an instantaneous view of a wave. The amplitude of the oscillations is shown on the y-axis, while the horizontal position of the wave is frozen on the x-axis.

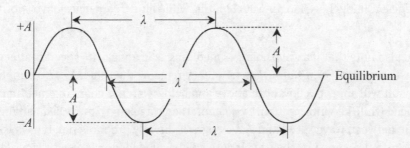

FIGURE 15.2 Principal wave components

The **amplitude**, A, is the maximum displacement, $A = x_{max}$. It can be measured from equilibrium ($x = 0$) to either $+A$ or $-A$. The wave illustrated in Figure 15.2 is a sine wave because the wave begins at the origin and then proceeds upward toward the positive maximum displacement, $+A$. However, the wave might start at the positive maximum displacement (making it a cosine wave), at the origin and proceed downward, or at the negative maximum displacement and proceed upward. Regardless, the wave will exhibit a repeated sinusoidal pattern between the two maximum displacements.

When the wave pattern repeats, it is known as a **wavelength**, λ. Wavelength is the distance measured between two successive identical portions of a wave. It is often easiest to see the wavelength between two successive crests or two successive troughs. However, the wavelength can also be determined as the distance between three crossings of the equilibrium line.

Medium and Wave Speed

The **medium** is the substance through which a wave propagates. For example, sound waves most often move in the medium air. Motion of a wave through one, and only one, medium is at constant speed. When a wave changes medium, the wave speed most often changes to a new constant speed in the new medium. The **speed of the wave**, v, in a specific medium is the product of the wave's frequency, f, and its wavelength, λ. In addition, the constant-speed formula is also applicable when waves travel in a constant medium.

$$v = f\lambda \qquad \text{and} \qquad v = \frac{d}{t}$$

As long as the wave travels in only one medium, the wave speed will remain constant and the frequency and wavelength will vary inversely.

An example of this phenomenon is sound waves. Frequency of sound is perceived as **pitch**. As the frequency increases, the sound wave itself does not travel any faster. If increasing frequency did affect the wave speed, then high-pitched notes would reach the ears of an observer before low-pitched notes. This, however, is not the case. An observer detects multiple frequencies produced by a single source simultaneously. This means that high- and low-pitched sounds must have different wavelengths. Therefore, frequency and wavelength must vary inversely. High-pitched (high-frequency) sounds have short wavelengths. Low-pitched (low-frequency) sounds have long wavelengths.

Another example of this phenomenon is light waves. The frequency of light is perceived as color. Low-frequency light is perceived as red, and high-frequency light is perceived as blue. If increasing frequency did affect the wave speed, then different colors would arrive at the eye of an observer at different times. This is not the case. Red and blue frequencies, for example, produced by a single source are viewed simultaneously by an observer and appear magenta. Red frequencies simply have longer wavelengths, while blue frequencies have shorter wavelengths.

You should also note that if a wave changes mediums as it travels, its speed is affected. However, when a wave changes medium, the frequency remains the same. As an example, a yellow swimming suit will appear yellow both above and below water. If changing mediums changes wave speed while keeping frequency constant, then wavelength must be changing. The relationship between wave speed and wavelength is directly proportional. If a wave speeds up when entering a new medium, the wavelength will become longer. If a wave slows down when entering a new medium, the wavelength will become shorter.

Effect of Amplitude on a Wave

The amplitude of a wave is the maximum displacement of the oscillating particles composing the wave as measured from their equilibrium positions. The amplitude of a wave does not affect the wave speed, wavelength, or frequency of the wave. The amplitude affects only the energy of the wave. A good example of this is the effect of amplitude on a sound wave. In terms of sound, amplitude is the volume of a sound. If, for example, a note of a certain frequency is being played through a loudspeaker, the note will not change if the volume is increased. Similarly, if the amplitude (volume) is increased, the speed of the sound coming out of the loudspeaker will not travel any faster to its intended observer.

Mechanical Waves

Mechanical waves involve the displacement of molecules in a medium from a position of equilibrium. Examples of mechanical waves and their mediums include vibrations on a string, ripples on a pond, and sound moving through air. A medium can therefore be a solid, a liquid, or a gas. These waves are disturbances created by a source that travels outwardly from the source at a constant velocity for that medium.

Sound Waves

Sound is a form of mechanical waves that travels as longitudinal (compression) waves. Longitudinal sound waves are difficult to illustrate. However, they mathematically graph as sinusoidal functions. As a result, graphs of sound waves often make them appear similar to transverse waves for illustrative purposes. Mechanical waves, such as sound, require a medium in order to propagate and travel. Sound can travel through solids, liquids, and gases. Each medium affects the speed and wavelength of the sound but not the frequency. The speed of a wave in a medium depends on many factors (density, elasticity, temperature, etc.). However, as a general rule, sound travels faster in denser mediums (see Table 15.2). This means sound moves fastest in solids and slowest in gases.

TABLE 15.2 Speed of Sound in Different Mediums

Medium	Speed of Sound
Aluminum	5,100 m/s
Water	1,480 m/s
Air (25°C)	345 m/s

Electromagnetic Waves

Electromagnetic waves are **light waves**. These include radio waves, microwaves, infrared light, visible light, ultraviolet light, X rays, and gamma rays. Researchers determined in the early twentieth century that these waves do not require a medium in which to travel and can therefore move through the vacuum of space. However, if they do travel in a medium, such as air, water, or glass, their wave behavior will be affected by that medium.

All forms of electromagnetic waves travel at the same constant speed in a vacuum. This value is known as the speed of light, c. Its value is 3×10^8 meters per second. For light moving in a vacuum, the speed formula can be modified.

$$c = f\lambda$$

The speed of electromagnetic waves is the complete opposite of mechanical waves. Electromagnetic waves have their highest speed in a vacuum, where mechanical waves cannot even exist. While mechanical waves typically speed up in denser mediums, electromagnetic waves are slowed as a medium's optical density increases. Table 15.3 shows commonly encountered speeds for light waves.

TABLE 15.3 Speed of Light in Different Mediums

Medium	Speed of Light (approximate)
Air/Vacuum	3.00×10^8 m/s
Water	2.25×10^8 m/s
Glass	2.00×10^8 m/s

A practical example of this phenomenon is seen when a beam of light from a laser pointer strikes a glass block at an angle. The light beam will move through the glass at a new angle due to the change in speed of light when it enters the glass block. This phenomenon, known as refraction, will be covered later.

Electromagnetic Spectrum

The oscillations of an electromagnetic field create a spectrum of electromagnetic waves. These waves transmit energy in direct proportion to their frequency. At the low end of the spectrum are radio waves. At the high end are X rays and gamma rays. In between these ends of the spectrum is visible light. The human eye is sensitive to the frequencies in this range, which are perceived as colors. The electromagnetic spectrum is shown in Figure 15.3. It starts at the left with long-wavelength, low-frequency, and low-energy radio waves. The spectrum progresses in order to the short-wavelength, high-frequency, high-energy gamma rays. Visible light is broken into the colors of the rainbow in their relative order as well.

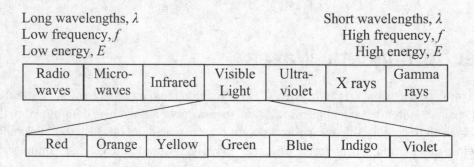

FIGURE 15.3 The electromagnetic spectrum

Doppler Effect

Wave Front Model of Sound

Diagrams of sound waves often show the source of sound as a dot and the waves of sound as expanding circles, similar to the result seen when dropping a rock into a pond. When a rock is dropped into a still pond, waves move outward in every direction from the point of impact.

The crests of the waves appear as expanding circles when viewed from above. This view of waves is known as the **wave front model**. The expanding circles represent the expanding wave crests. The distance between the circles (crests) is the wavelength. Although drawn as circles on paper, sound-wave fronts actually form expanding three-dimensional spheres. A stationary sound source would emit spherical wave fronts as depicted in Figure 15.4.

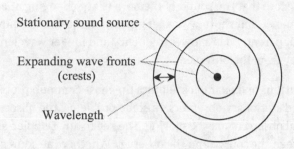

Stationary sound source

Expanding wave fronts
(crests)

Wavelength

FIGURE 15.4 Wave front model

The Doppler Effect and Sound

When a wave is produced by a source in a medium, the wave propagates through the medium at a constant speed, v. Once produced by the source, the speed of the wave in that particular medium will not change, regardless of the speed of the source.

An interesting phenomenon occurs when the source of a sound is moving with respect to a stationary observer. Figure 15.5 shows a car moving to the right at constant speed. If the driver presses the horn continuously, then sound waves will leave the car and travel outward in expanding spheres.

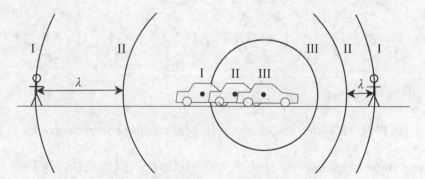

FIGURE 15.5 The Doppler effect

To make it simple, assume one sound wave is emitted every second. Each sound-wave front moves outward from the location where the car was at the time the wave was emitted. The first sound wave, I, was emitted when the car was at position I. This wave has been expanding for three seconds. The second wave, II, was emitted a second later when the car was at position II. This wave has been moving for two seconds. The third wave, III, was emitted when the car was at position III. This wave has been traveling for only one second.

The sound of the horn will differ depending on the location of an observer, and the effects observed are known as the **Doppler effect**. For an observer in front of the car, the wavelengths appear to be shorter than they actually are. Since the speed of sound is constant, the shorter wavelengths create a higher frequency, and the horn will sound as though it has a higher pitch than it actually does. For an observer behind the car, the wavelengths appear to be longer than they actually are. The observer hears a lower frequency with a lower pitch than the horn actually makes. Note that the sound of the horn is not changing at all. It merely appears to have a higher frequency and shorter wavelengths when the sound source is moving toward the observer, It also apparently has a lower frequency and longer wavelength when the sound source is moving away from the observer.

The previous explanations and examples address the most common problems involving moving sound sources and stationary observers. In some problems, the observer may be moving, or both the source and observers may be moving. The key to any Doppler-shift problem is the relative motion between the source and the observer. Whether the source moves toward the observer, the observer moves toward the source, or they both move toward each other, the observed effect is the same. When a sound source and observer approach each other, the perceived wavelength of sound decreases and the observed frequency increases (high pitch). If the distance between the source and observer is increasing, then the perceived wavelength increases and the observed frequency decreases (low pitch).

Speed of the Source and the Speed of Sound

Several scenarios are possible depending on the speed of the source of sound. Figure 15.6 depicts four possible Doppler-effect diagrams for moving sound sources.

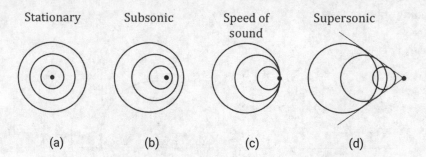

Stationary	Subsonic	Speed of sound	Supersonic
(a)	(b)	(c)	(d)

FIGURE 15.6 The Doppler effect for moving sound sources

A **stationary object** will create concentric circular wave fronts as seen in Figure 15.6(a). A **subsonic object**, as seen in Figure 15.6(b), is an object moving with a speed less than the speed of sound (about 340 m/s in air). The best examples are a car continuously honking its horn or a siren on an emergency vehicle. When the sound source reaches the **speed of sound**, as shown in Figure 15.6(c), the wave fronts pile up on one another. These sound waves constructively interfere (add up) with each other to create a phenomenon known as the **sound barrier**. If a sound source exceeds the speed of sound, it is said to be **supersonic**, which is shown in Figure 15.6(d). When this occurs, the wave fronts constructively interfere in a manner that creates a wake of sound that is similar to the wake of a boat. When this wake of sound passes by a person, the compression of waves sounds like a boom. This is known as a **sonic boom**.

Superposition and Standing Waves

Superposition

Superposition is when two or more waves occupy the same point in space, at a given moment, and their combination displaces the medium to reflect the sum of their individual displacements. In Figure 15.7, two wave pulses, A and B, generated in a string travel toward one another. In Figure 15.7(a), the pulses are the same size and are on the same side of the string. In Figure 15.7(b), the pulses are the same size, but pulse B is inverted.

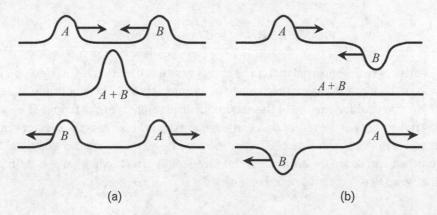

FIGURE 15.7 Constructive (a) and destructive (b) wave pulses

When pulses A and B superimpose (occupy the same location), their combined displacements add to create a composite pulse. After superimposing for an instant, the wave pulses continue on their way in their original directions. In Figure 15.7(a), the wave pulses add to create a larger wave. This is an example of **constructive interference**. The waves overlap. The medium displacement is larger than it was for the waves individually. Figure 15.7(b) is an example of **destructive interference**, which occurs when waves overlap and cause a smaller displacement of the medium. When opposite waves are exactly the same size, they will cancel entirely, as shown in Figure 15.7(b).

Standing Waves

When a wave is trapped between two boundaries, the individual points move up and down but do not travel. They appear to stay in one place. This is known as a **standing wave**. An example can be seen on a guitar in which the guitar strings are attached at one end to the headstock and at the other end to the bridge. Plucking the string will create a standing wave. As one wave strikes a boundary, it bounces back and interferes with the next incoming wave. This creates an alternating pattern where part of the standing wave moves back and forth while several points do not appear to move at all. Figure 15.8 shows the incoming wave as a solid line and the reflected wave as a dashed line.

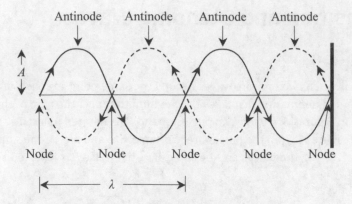

FIGURE 15.8 A standing-wave pattern

The **nodes** are where the superposition of two waves creates destructive interference. The **antinodes** are the locations of greatest constructive interference. The amplitude, A, is the distance from the equilibrium line to the maximum displacement of the string. The maximum amplitude occurs at the antinodes. The wavelength of the wave is the distance between three successive nodes or three successive antinodes. The nodes are spaced exactly $\lambda/2$ away from each other, and so are the antinodes. The wavelength of a standing wave is 2 times the distance between two successive nodes or antinodes.

Fundamental Frequency and Harmonics

When a simple instrument such as a guitar string or a flute is played, standing waves occur in the string or in the air inside the flute. The simplest standing waveform that can be produced in the string or between the ends of the flute produces a frequency known as the **fundamental frequency**, f_1. The fundamental frequency is also known as the **first harmonic**. In a string and in an open tube (open at both ends), the fundamental frequency is associated with a waveform consisting of half of a wavelength. For closed tubes (closed at one end), the waveform consists of a quarter wavelength. Figure 15.9 shows the standing waveform associated with each of these simple instruments.

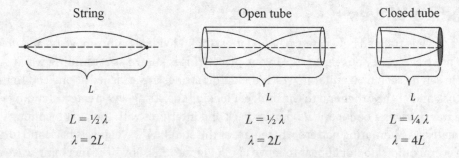

FIGURE 15.9 First-harmonic waveforms

The values for wavelengths determined in Figure 15.9 can now be substituted into the wave-speed equation in order to determine the fundamental frequency, f_1.

- Strings and open tubes: $v = f_1 (2L)$
- Closed tubes: $v = f_1 (4L)$

Other standing waves can be created in strings and in tubes. All of these patterns are known as the **harmonics**. They have a variety of frequencies (f_1, f_2, etc.) associated with the number of wavelengths fitting along the string or between the ends of the tubes. Usually problems about the harmonics are associated with the difference in frequency between the fundamental frequency, f_1 (the first harmonic), and the second harmonic, f_2. More than likely, these problems will focus on either a string or an open tube, which are mathematically the same. To create the second harmonic, the next complete standing wave must fit along the string or inside an open tube. Figure 15.10 shows the next wavelength capable of creating a standing wave in strings and tubes. The wavelengths depicted are associated with the frequencies of the second harmonic, f_2.

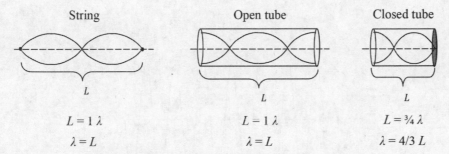

FIGURE 15.10 Second-harmonic waveforms

For strings and open tubes, the wavelength for the second harmonic is half of the wavelength associated with the fundamental. The wavelength of the harmonics can be easily determined if the wavelength, λ_1, associated with the fundamental frequency is known. Simply multiply this key wavelength by the inverse of the harmonics number, n.

$$\lambda_n = (1/n)\lambda_1$$

The wavelength of the second harmonic can be calculated as follows:

$$\lambda_2 = (1/2)\lambda_1$$

Frequency and wavelength are inversely proportional. As a result, the frequencies can be found by multiplying the fundamental frequency by n.

$$f_n = nf_1$$

Beats

So far, the examples of superposition are for waves traveling along the same medium at the same frequency and wavelength. However, there can also be superposition of waves that do not have the same frequency. When this occurs, the resulting superposition does not reflect a perfect sinusoidal pattern. A good example of this is the superposition of sound waves produced by two musical instruments at slightly different pitches (musical notes).

Imagine two violinists playing the same note. A sound wave is produced from each violin. As the waves propagate through the medium of air, they will undergo superposition. If the waves are perfectly identical, they will complement each other and add constructively. If, however, the waves are slightly different, there will be both constructive and destructive portions. The destructive portions decrease the amplitude (volume) at a regular rate and cause what are known as **beats**.

In Figure 15.11, two completely different waves are added to create a third wave pattern, which displays the characteristic known as beats. Figures 15.11(a) and 15.11(b) display the amplitude versus time graphs for two sound waves, which have different frequencies and wavelengths. Figure 15.11(c) illustrates the superposition of these two waves.

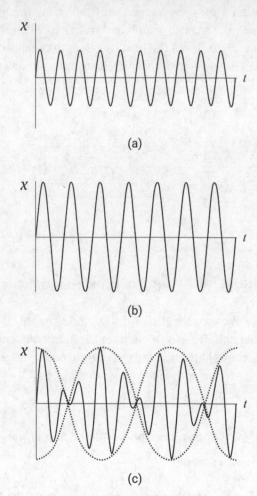

(a)

(b)

(c)

FIGURE 15.11 Beats

Three distinct beats occur during the graphed time interval. This is known as the **beat frequency**. The beat frequency created by any two waves can be quickly determined by the absolute value of the difference between their frequencies.

$$f_{\text{beat}} = |f_1 - f_2|$$

Beats are detected by the human ear as quick outbursts of loud and quiet because amplitude (volume) is being affected. Most people can detect seven or fewer beats per second. For two waves that are very dissimilar in frequency, there are too many beats per second to be detected by the human ear.

Geometric Optics

In this chapter, you will learn how to:

- Predict the location and appearance of images due to reflection
- Analyze refraction, the bending of light rays when they move from one medium into another
- Determine the location of and identify the types of images formed by spherical lenses and mirrors

When light strikes a new medium, such as a reflective mirror, transparent glass, or water, a variety of things can occur. Light can be absorbed, reflected, or refracted (bent) when it strikes the new medium. Usually, all of these occur to some degree. For simplicity, however, problems often focus on one of these possibilities at a time.

Table 16.1 lists the variables that will be used.

TABLE 16.1 Variables for Geometric Optics

New Variables	Units
n = index of refraction	No units
θ_c = critical angle	Degrees
R = radius of curvature	m (meters)
f = focal distance	m (meters)
d_o = object distance	m (meters)
d_i = image distance	m (meters)
h_o = object height	m (meters)
h_i = image height	m (meters)
M = magnification	No units but may be followed by a ×

Ray Model of Light

Light can be viewed in many ways. Each model of light has its advantages in solving specific problems. In certain problems, it is advantageous to view light as advancing wave fronts (crests). The angles at which light strikes a surface will be important. So, a different model of light is needed. The three models of light are represented visually in Figure 16.1.

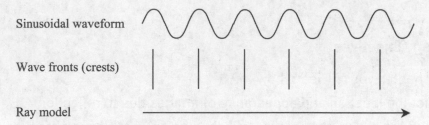

FIGURE 16.1 Three models of light

The **ray model** of light shows the path of light as a straight line with an arrow indicating the direction of the light. The advantage of using the ray model of light is it allows angles of reflection and refraction to be measured.

Reflection

When light strikes a flat, polished surface, it is reflected in a manner similar to a ball's bouncing off of a wall or the floor. In Figure 16.2, an **incident ray** (inbound ray) of light is shown striking a surface at an **incident angle** of θ_i. Whenever surfaces are involved, angles are always measured from a **normal** line. The normal is a line drawn perpendicular to the surface area. The resulting **reflected ray** bounces off the surface with an **angle of reflection** of θ_r.

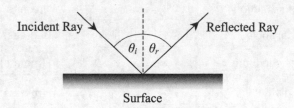

FIGURE 16.2 Reflection

The angle of reflection is the same as the incident angle. This relationship is known as the **law of reflection**.

$$\theta_i = \theta_r$$

Specular reflection occurs when a surface is flat, smooth, and polished; all the reflected rays leave the surface parallel to one another, and the image of the reflection appears similar to the original object. Rough surfaces create **diffuse reflection**, where the irregularities of these surfaces cause the reflected rays to move off in random directions. Reflection allows us to see these rough surfaces, but clear images of objects cannot be reflected by rough surfaces.

Plane Mirror

The simplest mirror is a flat mirror, known as a **plane mirror**. When an object, such as a person, is viewed in a plane mirror, several rays of light can be visualized as starting from the object, moving toward the mirror, and reflecting off of the mirror's surface. Each ray follows the law of reflection. Images are formed where rays intersect. However, the rays starting from the head of the person in Figure 16.3 diverge (spread out). If the reflected rays are traced backward (back trace), shown as dashed lines in Figure 16.3, then an intersection can be found. The image will appear at the intersection of reflected rays.

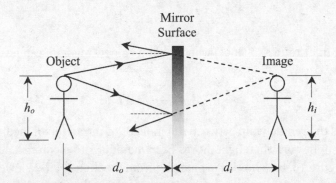

FIGURE 16.3 Reflection off of a plane mirror

The image formed by a plane mirror has several characteristics.

1.   The image is upright. It appears right side up.
2.   The image is the same size as the object, $h_i = h_o$, resulting in a magnification equal to one, $M = 1$.
3.   The image distance and object distance are equal, $d_i = d_o$.
4.   The image is a **virtual image**. No light passes through the mirror. Therefore, intersecting light does not form the image. This means the image cannot be projected onto a screen, which is an important characteristic of a virtual image. Yet virtual images can be seen. An image is easily recognized as a virtual image since it is always upright.
5.   Plane mirrors reverse front and back.

Refraction

When light moves into a new medium with a different density, its speed and wavelength change. If the light strikes this medium at an angle, the ray will bend at the surface of the new medium. The bending of light due to a change in density is known as **refraction**.

Index of Refraction

The **index of refraction**, n, is a ratio of the speed of light in a vacuum, c, to the speed of light in the medium, v, in which the light is moving.

$$n = \frac{c}{v}$$

Since both c and v are measured in meters per second, the units cancel, and the index is simply a value with no units. Light moves the fastest in a vacuum and is slower in a medium. This means the numerator, c, will either be equal to or greater than the denominator, v. As a result, the index of refraction can never be less than 1. When light is traveling in a vacuum, v and c have the same value, and the index of refraction is equal to 1, $n_{vacuum} = 1$. Air has a very low density. The speed of light in air is very nearly equal to the speed of light in a vacuum. This means that the accepted value for the index of refraction of air is also 1, $n_{air} = 1$. Knowing that the indexes of refraction for a vacuum and for air are both equal to 1 is often important when solving problems. The index of refraction increases as optical density increases.

Snell's Law

The amount that light refracts (bends) when it enters a new medium can be determined using **Snell's law**.

$$n_1 \sin \theta_1 = n_2 \sin \theta_2$$

Snell's law shows the relationship between the indexes of refraction of mediums 1 and 2 and the angles of the light rays when light moves from medium 1 to medium 2. Figure 16.4 shows two examples. Figure 16.4(a) shows a ray moving from medium 1 with a lower optical density into medium 2 with a higher optical density. The light ray is bent toward the dashed normal line. Figure 16.4(b) on the right shows a ray moving from medium 1 with a higher optical density into medium 2 with a lower optical density. This time the light ray bends away from the dashed normal line. In both diagrams, the larger angle is found in the medium that has the lower optical density and the lower index of refraction. The smaller angle is located in the optically denser medium with the higher index of refraction.

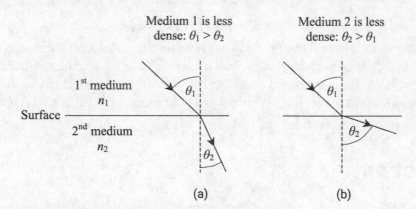

FIGURE 16.4 Snell's law of refraction

Refraction has some specific characteristics that are worth noting.

1. Refraction occurs only if the two mediums have different optical densities and different indexes of refraction. If the indexes of refraction are equal, no refraction occurs even though the light travels through two different mediums.

2. In order for light to refract, it must strike the boundary between the two mediums at an angle that is not perpendicular to the surface. If light hits the boundary perfectly

perpendicular to the surface (the light ray follows the normal), then $\theta_1 = 0°$ and no bending due to refraction occurs. In this case, the light will continue along the normal line without bending, $\theta_2 = 0°$. The speed of the light will change upon entering the new medium. However, the effect will not be visibly noticeable as the angle with reference to the normal line is zero.

3. The optically denser medium will have the larger index of refraction, n, the slower speed of light, v, the shorter wavelength, λ, and the smaller angle, θ.

Thin Lenses

Lenses use refraction to change the way an object appears. When light traveling in air enters the denser lens, it slows and bends due to refraction. When it exits the lens and moves back into air, the light speeds up and experiences a second refraction. The type and degree of curvature of the two lens surfaces dictate where the image will appear and how it will be magnified.

Beginning physics introduces the geometric optics of extremely simplified lenses. The surfaces of these lenses are spherical. As a result, they are not perfect. The images formed by spherical lenses have a slight aberration, causing them to lack sharpness. The spheres forming these lenses are the same size, creating symmetrical lenses. In addition, the lenses are assumed to be extremely thin even though they may not appear very thin in diagrams. These assumptions greatly simplify lens mathematics.

Figure 16.5 demonstrates the symmetrical and spherical nature of a simple **convex lens**. Although the diagram appears to be two circles, the lens is actually formed by the intersection of the two spheres. The optical axis is a horizontal line running through the center of the lens. All vertical measurements are made from the optical axis. A vertical line (usually not drawn) also passes through the center of the lens. All horizontal measurements are made from the center of the lens. Two key points, f and $2f$, are shown on both sides of the lens. The symbols f and $2f$ also represent distances measured from the center of the lens to these key points. The point $2f$ is at the center of curvature of each sphere. The distance $2f$ is, therefore, equal to the radius, R. As a result, the focal length, f, can be calculated as follows:

$$2f = R \qquad \text{rearranges to} \qquad f = \frac{R}{2}$$

This means that the focal point, f, is located midway between the center of curvature and the surface of the lens. The lenses are considered so thin that they have negligible thickness (despite the obvious thickness shown in lens diagrams). As a result, the focal distance is measured from the center of the lens rather than the lens surface. Although only a convex lens is shown in Figure 16.5, the focal-length formula holds true for all spherical lenses and mirrors discussed.

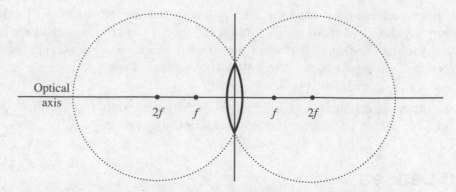

FIGURE 16.5 A simple convex lens

Converging Lenses, Object Outside *f*

Converging lenses bring light together. The convex lens is a converging lens. The location and size of the image formed by a lens or mirror can be plotted using ray tracing. In ray tracing, an object is represented by a vertical arrow extending upward from the optical axis, as shown in Figure 16.6. Two different light rays, both starting at the top of the object (tip of the vertical arrow), are drawn moving through the lens. Where they intersect is the location where the image will be focused. Look at the figure and trace the following key rays.

- Light parallel to the optical axis converges on the far focal point.
- Light that passes through the center of the lens moves in a straight line.
- Light passing through the focus exits parallel to the optical axis.

Figure 16.6 shows these three key rays of light originating at the object and intersecting at the image formed by this converging convex lens.

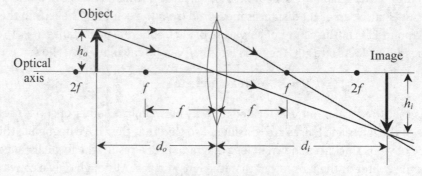

FIGURE 16.6 Convex lens

The rules for positive and negative signs for d_o, d_i, h_o, and h_i are the same as those for the pinhole camera, discussed earlier. There is one additional sign to consider. Converging optical instruments, such as the convex lens in Figure 16.6, have a positive focal distance $(+f)$. When the object is positioned outside of the focal point, f, the image will be inverted and real. For example, movies are projected on a screen. Therefore, they are real images and inverted. In order to show a movie upright in the theater, the film must be fed into the projector inverted.

There is a key geometric relationship between the focal length, f, of the lens, the object distance, d_o, and the image distance, d_i.

$$\frac{1}{f} = \frac{1}{d_o} + \frac{1}{d_i}$$

All of these variables are in the denominator and require a value to be inverted during the solution process. For example, when solving for f, the answer for $1/f$ will be one of the available choices. After solving for $1/f$, remember to invert to find f.

$$M = \frac{h_i}{h_o} = -\frac{d_i}{d_o}$$

Table 16.2 summarizes the signs for the variables in Figure 16.8 and in the equations shown above.

TABLE 16.2 A Converging Convex Lens with an Object Outside of f

If You See . . .	Key Fact	Result
Converging convex lens	Converging	$+f$
Object arrow points upward	Upright object	$+h_o$
Image arrow points downward	Inverted image	$-h_i$
The sign on magnification matches h_i	Inverted image	$-M$
Distance to the object is always positive	Always	$+d_o$
The sign on image distance is opposite h_i	Inverted image	$+d_i$
If $-h_i$ or $+d_i$, the image is real If $+h_i$ or $-d_i$, the image is virtual	$-h_i$ or $+d_i$	Real
Light transmits through lenses to the far side	Light forms real images	Far side

Problems may simply address trends as the object is moved either toward the focal point or away from the focal point. For the converging convex lens, remember the following points:

1. Object outside of $2f$: small image ($M < 1$) and inside $2f$ on the far side.
2. Object at $2f$: image and object are the same size ($M = 1$) and at $2f$ on the far side.
3. Object between $2f$ and f: large image ($M > 1$) and outside $2f$ on the far side.
4. As objects move toward f, the image distance and the image size increase.

Converging Lenses, Object at f

When the object is placed at the focal point of a converging lens, the ray traces are parallel to each other and never intersect. No image is formed. However, if this scenario is reversed and the object is positioned infinitely far away, then all the light rays arriving at the lens will be parallel to the optical axis. Every ray of light will converge at the focal point, creating an image at f. Wavelengths of light are so incredibly small that in comparison, an object 100 meters away might as well be at infinity. You should know that the image of a distant object will be located at the focal point on the opposite side of a converging convex lens.

Converging Lenses, Object Inside *f*

An interesting phenomenon occurs when the object is moved inside the focal point of a converging convex lens. In Figure 16.7, the ray traces used in Figure 16.6 do not intersect on the far side of the lens. However, if they are traced backward (dashed lines), an intersection is found on the near side of the lens. This creates an upright image, which is a virtual image that cannot be projected onto a screen. The formulas used previously remain the same. However, the signs on the upright image height ($+h_i$) and image distance ($-d_i$) reverse when the object moves inside the focal point. Even though the ray traces seem to diverge, this is still a converging convex lens with a positive focal point ($+f$). The actual light rays are still being converged on, and pass through, the focal point. The image will have the greatest magnification when the object is closest to the focal point. As the object is moved from the focal point, *f*, toward the lens, the image decreases in size and moves toward the lens. An example of this is the magnifying lens. An object placed between the focal point and magnifying lens will be upright, virtual, and magnified.

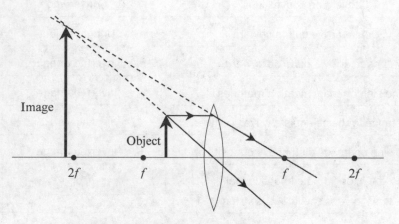

FIGURE 16.7 Magnifying convex lens

Table 16.3 summarizes the signs for the variables in Figure 16.7.

TABLE 16.3 A Converging Convex Lens with an Object Inside of *f*

If You See . . .	Key Fact	Result
Converging convex lens	Converging	$+f$
Object arrow points upward	Upright object	$+h_o$
Image arrow points upward	Upright image	$+h_i$
The sign on magnification matches h_i	Upright image	$+M$
Distance to the object is always positive	Always	$+d_o$
The sign on image distance is opposite h_i	Upright image	$-d_i$
If $-h_i$ or $+d_i$ the image is real If $+h_i$ or $-d_i$ the image is virtual	$+h_i$ or $-d_i$	Virtual
The rays trace back to form a virtual image	Opposite side of light	Near side

EXAMPLE 16.1

Convex Lenses

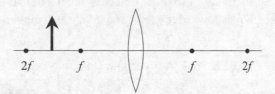

The object viewed by a convex lens is positioned outside of the focus, as shown in the diagram above. Which of the following correctly describes the image?

(A) No image is formed
(B) Real and upright
(C) Real and inverted
(D) Virtual and upright
(E) Virtual and inverted

WHAT'S THE TRICK?

Convex lenses produce three possible outcomes. When the object is inside of f, the image is virtual and upright. When the object is at f, the image cannot be formed. When the object is outside of f, the image is real and inverted, which matches the scenario given in the problem. The answer is C.

Diverging Lenses

The **concave lens** is a **diverging lens**. Diverging optical instruments spread rays of light. Instead of the light converging on the far focus, it spreads out from the near focus, as shown in Figure 16.8 and summarized as follows:

- Light parallel to the optic axis diverges in a line from the near focal point.
- Light that passes through the center of the lens moves in a straight line.
- The resulting rays diverge, and their back traces intersect. (Note that the back trace of the straight ray passing through the center of the lens coincides with the ray itself.)

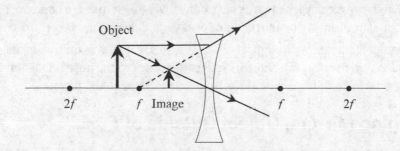

FIGURE 16.8 Concave lens

The resulting image is upright and virtual. They are summarized in Table 16.4. Unlike the converging lens, the diverging concave lens is capable of producing only a small ($M < 1$) image that is upright and virtual. Whether the object is outside of f, at f, or inside of f has no bearing. As the object moves toward the lens, the image also moves toward the lens and becomes larger.

TABLE 16.4 A Diverging Concave Lens

If You See . . .	Key Fact	Result
Diverging concave lens	Diverging	$-f$
Object arrow points upward	Upright object	$+h_o$
Image arrow points upward	Upright image	$+h_i$
The sign on magnification matches h_i	Upright image	$+M$
Distance to the object is always positive	Always	$+d_o$
The sign on image distance is opposite h_i	Upright image	$-d_i$
If $-h_i$ or $+d_i$, the image is real If $+h_i$ or $-d_i$, the image is virtual	$+h_i$ or $-d_i$	Virtual
The rays trace back to form a virtual image	Opposite side of light	Near side

Spherical Mirrors

Spherical mirrors are simply small sections of a single sphere that have a reflective surface. For **concave mirrors**, the reflective surface is the inside surface. For **convex mirrors**, it is the outside surface. Mirror-optics problems are nearly identical to lens problems. The equations are the same. The rules for the variable signs are the same. The trends of image location are the same. However, there are a few key differences.

1. Although lenses consist of two intersecting spheres with two focal points, mirrors consist of a single spherical surface with only one focal point.

2. Although converging lenses are convex and diverging lenses are concave, mirrors are the opposite. Converging mirrors are concave, and diverging mirrors are convex.

3. Although light passes through a lens and creates real images on the far side and virtual images on the near side, mirrors reflect light back to the near side. For mirrors, real images form on the near side and virtual images form on the far side.

Converging Mirrors, Object Outside of f

The concave mirror converges parallel rays of light through the focal point. Ray tracing for mirrors requires you to use a different strategy than that for a lens.

- Light parallel to the optical axis is reflected through the focus.
- Light through the focus reflects parallel to the optical axis.

The ray trace in Figure 16.9 demonstrates how the image can be found when the object is located outside the focus.

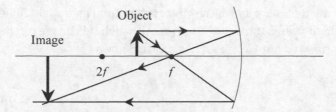

FIGURE 16.9 Concave mirror

The mathematics, signs, and trends shown in Table 16.5 are identical to those of the converging lens with an object outside its focus.

TABLE 16.5 Converging Mirrors with the Object Outside of *f*

If You See . . .	Key Fact	Result
Converging concave mirrors	Converging	$+f$
Object arrow points upward	Upright object	$+h_o$
Image arrow points downward	Inverted image	$-h_i$
The sign on magnification matches h_i	Inverted image	$-M$
Distance to the object is always positive	Always	$+d_o$
The sign on image distance is opposite h_i	Inverted image	$+d_i$
If $-h_i$ or $+d_i$, the image is real If $+h_i$ or $-d_i$, the image is virtual	$-h_i$ or $+d_i$	Real
Light reflects off of mirrors to the near side	Light forms real images	Near side

Converging Mirrors, Object at *f*

This is the same as for a converging lens. When the object is positioned at the focus, no image is formed since the light rays are parallel and cannot intersect. However, if the object is positioned far away, the light rays arriving at the mirror will be essentially parallel. Parallel rays striking a converging optical instrument are refracted and focused at the focal point, *f*. A practical example of this is the collection of light from a distant star through a concave, reflecting telescope.

Converging Mirrors, Object Inside of *f*

Just as with the converging lens, when the object is moved inside the focus, an upright, virtual image is created. The ray-trace rules are a bit more complicated for this scenario and are depicted in Figure 16.10. The ray parallel to the optical axis reflects through the focus as before. However, a ray drawn from the tip of the object through the focus will not strike the mirror. A new ray must be drawn. The mirror is spherical, and 2*f* is at the center of the sphere. Any ray of light starting at the center of a sphere, 2*f*, will be reflected straight back to the center, 2*f*. A ray of light starting at 2*f* and passing through the very tip of the object will reflect right back toward 2*f*, as shown in Figure 16.12. This ray diverges from the ray passing through *f*.

When back ray traces are drawn they intersect to form an upright image on the far side of the mirror. No light passes through the mirror. Although this image can be seen with the eye, it cannot be projected onto a screen. It is a virtual image.

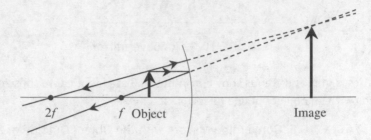

FIGURE 16.10 Virtual image in a concave mirror

All the equations, variable signs, and image trends are identical to those seen in the converging lens when the object is located inside of f. The main difference is that virtual images form on the far side of mirrors. Table 16.6 summarizes this information.

TABLE 16.6 Converging Mirrors with the Object Inside of f

If You See . . .	Key Fact	Result
Converging concave mirrors	Converging	$+f$
Object arrow points upward	Upright object	$+h_o$
Image arrow points upward	Upright image	$+h_i$
The sign on magnification matches h_i	Upright image	$+M$
Distance to the object is always positive	Always	$+d_o$
Sign on image distance is opposite h_i	Upright image	$-d_i$
If $-h_i$ or $+d_i$, the image is real If $+h_i$ or $-d_i$, the image is virtual	$+h_i$ or $-d_i$	Virtual
The rays trace back to form a virtual image	Opposite side of light	Far side

Diverging Mirrors

The diverging mirror is very similar to the diverging lens. It is convex rather than concave. The ray trace in Figure 16.11 appears very different, but it uses the same logic as seen earlier.

1. Light parallel to the optical axis reflects in a line drawn from the focus.
2. Light aimed at the focus reflects parallel to the optical axis.
3. The reflected rays diverge so the image is formed by their back traces.

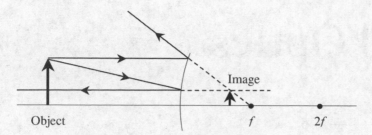

FIGURE 16.11 Diverging convex mirror

The resulting image is upright and virtual. It is very similar to the image seen in the diverging lens. The image is always small ($M < 1$), upright, and virtual. As the object moves toward the mirror, the image becomes larger and moves toward the mirror. The main difference is that this virtual image is on the far side of the mirror. Table 16.7 lists the characteristics of a diverging mirror.

TABLE 16.7 Characteristics of a Diverging Mirror

If You See . . .	Key Fact	Result
Diverging convex mirrors	Diverging	$-f$
Object arrow points upward	Upright object	$+h_o$
Image arrow points upward	Upright image	$+h_i$
The sign on magnification matches h_i	Upright image	$+M$
Distance to the object is always positive	Always	$+d_o$
The sign on image distance is opposite h_i	Upright image	$-d_i$
If $-h_i$ or $+d_i$, the image is real If $+h_i$ or $-d_i$, the image is virtual	$+h_i$ or $-d_i$	Virtual
The rays trace back to form a virtual image	Opposite side of light	Far side

CHAPTER 17

Physical Optics

Learning Objectives

In this chapter, you will learn how to:

- Explain the nature of diffraction and identify its characteristics
- Analyze the double-slit interference pattern
- Examine the polarization of light
- Explain why objects appear as a specific color and how light can be dispersed or scattered

Physical optics involves such topics as diffraction, interference, and polarization. These effects are usually analyzed using either the wave-front model of light or the sinusoidal form.

Table 17.1 lists the variables used.

TABLE 17.1 Variables for Physical Optics

New Variables	Units
m = a specific maximum or minimum	No units
x_m = distance measured to m above	m (meters)
θ_m = angle measured to m above	Degrees
d = slit spacing	m (meters)
L = distance from slits to screen	m (meters)

Diffraction

When waves pass near a barrier or through an opening (slit), they bend and spread out to fill the space behind the barrier or the slit. The bending of a wave due to a barrier or opening is known as **diffraction**. The amount of bending has to do with the size of the obstacle or opening compared with the wavelength of the wave. In the diagrams in Figure 17.1, parallel wave fronts are shown approaching various openings. The bending due to diffraction increases as the openings become smaller.

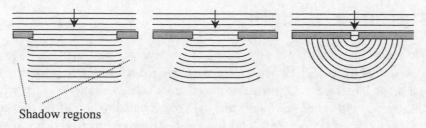

Shadow regions

FIGURE 17.1 Diffraction

When the opening is large as compared with the wavelength of the waves, the waves move through the opening with little diffraction. This creates large shadow regions with no wave activity to the left and right of the opening. When light waves are diffracted, the absence of light in the shadow regions leaves these areas dark. However, when the size of the opening is similar to the size of the wavelength of the waves, the diffraction is so pronounced that the spreading wave fronts form a circular pattern with no shadow regions. Since light is composed of very small wavelengths, openings that cause significant diffraction must be incredibly narrow.

A geometric explanation for the circular nature of the resulting diffraction pattern was proposed by Christian Huygens and is known as the **Huygens' principle**. There are two main aspects of his principle.

1. Every oscillator in a wave creates spherical wavelets that propagate outward.
2. The wave front created by these oscillators is due to the combined interference of the wavelets.

Figure 17.2 shows a diagram of a portion of a linear wave. The dots lying along the crest of the wave represent the individual oscillators, each creating circular wave fronts.

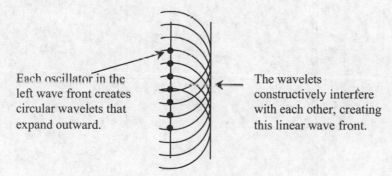

Each oscillator in the left wave front creates circular wavelets that expand outward.

The wavelets constructively interfere with each other, creating this linear wave front.

FIGURE 17.2 Wave fronts

This provides an explanation for the circular wave fronts seen when waves move through narrow openings. If the opening is very small, only one, or very few, oscillator(s) propagate(s) the wave through the opening. As a result, circular wave fronts are generated on the other side of the opening.

Interference of Light

When light waves interact, they can interfere constructively or destructively. If identical wave crests (represented by wave fronts) meet, **constructive interference** adds them together to create a single larger wave. Constructive interference causes water waves to increase in height. It causes sound to become louder and light to become brighter. If, on the other hand, a wave crest meets a wave trough of identical size, **destructive interference** will cancel these waves entirely. This creates an absence of wave activity. It causes water to be flat, quiet instead of loudness, and darkness instead of brightness.

Young's Double-Slit Experiment

Thomas Young constructed an experiment, known as **Young's double-slit experiment**, involving the interference of light. He shined **monochromatic light**, which is light composed of only one wavelength, on two incredibly narrow openings (slits) that were near each other. The light diffracted through each slit, causing circular wave fronts to spread outward. These patterns overlapped each other and created both constructive and destructive interference. The resulting pattern was projected onto a screen, where constructive interference created bright regions (bright fringes or maximums) and destructive interference created dark regions (dark fringes or minimums). Figure 17.3 shows the interference pattern created by the circular wave fronts spreading from each slit.

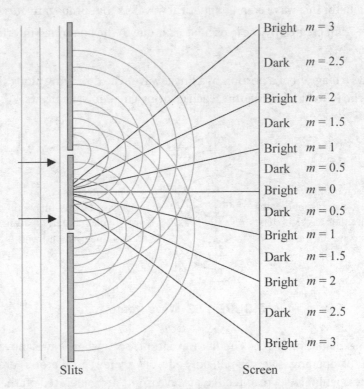

FIGURE 17.3 Young's double-slit experiment

The expanding circular wave fronts represent the wave crests. Where wave crests intersect, they interfere constructively. In Figure 17.3, lines through successive wave-crest intersections have been added. They extend from a point midway between the slits toward the screen. Where these lines of constructive interference hit the screen is where bright bands (maximums, *m*) of light will be seen. In the regions in between, destructive interference creates bands that are dark (minimums, *m*). The maximums and minimums are numbered starting with the central maximum, *m* = 0. The bright maximums are numbered as whole numbers (*m* = 1, 2, or 3 . . .), while the minimums are numbered with halves (*m* = 0.5, 1.5, or 2.5 . . .). Be very careful with the values assigned to dark minimums. The value for the third dark minimum (third dark fringe) is *m* = 2.5 and not 3.5 as students often incorrectly believe. The value representing the maximums and minimums has no units, and the reason behind the numbering system will be discussed below.

A mathematical relationship describes the resulting bright and dark fringes. This will be easier to explain if the diagram is simplified to solve for one specific maximum. Figure 17.4 shows the mathematical relationships for the first bright maximum, *m* = 1. Only the lines extending to the central (reference) maximum and the first maximum are shown.

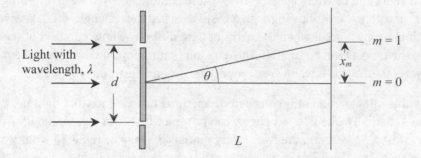

FIGURE 17.4 First bright fringe position for the double-slit experiment

Various key variables have been labeled in Figure 17.4. These include the number for the maximum to be analyzed, *m*; the distance from the central maximum to the maximum being investigated, x_m; the spacing between the slits, *d*; the distance from the slits to the screen, *L*; and the angle, θ. The actual widths of the slits themselves are not needed. The variable *m* has no units, and the angle is measured in degrees. All other variables are lengths measured in meters. Students often confuse *m* and x_m. Remember that *m* is the number for the maximum and x_m is the distance to that maximum. There are two mathematical relationships for Young's double-slit experiment.

$$x_m \approx \frac{m\lambda L}{d} \quad \text{and} \quad d\sin\theta = m\lambda$$

The formula on the left is actually an approximation that works well as long as the screen distance, *L*, is very large.

When the diagram is viewed differently, an explanation for the numerical values of *m* is apparent. In Figure 17.5, two rays of light are shown moving from each slit toward the maximum, *m*.

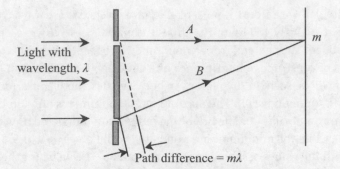

FIGURE 17.5 Path difference

Light ray *B* must travel a longer distance than light ray *A*. The difference between the distance the two rays travel is known as the **path difference**. Light reaching the first maximum, *m* = 1, has a path difference equal to 1 wavelength. Light reaching the second maximum, *m* = 2, has a path difference equal to 2 wavelengths. In other words, the maximum numbers, *m*, represent the number of wavelengths by which the paths differ. Bright maximums occur when crests meet and interfere constructively. At maximum *m* = 1, a wave following path *A* must meet up with a wave following path *B* that is exactly 1 wavelength off. Likewise, all the other maximums must occur when the paths of light differ by whole numbers of wavelengths. Otherwise, their crests will not match. The dark minimums appear when the waves are off by half of a wavelength. This is when crests meet troughs and destructively interfere.

Young's double-slit experiment is considered experimental evidence that light possesses a wave characteristic. The double-slit effects can be seen with water waves, and it can be heard with sound waves. Sometimes, the two closely spaced slits may be replaced with two closely spaced loudspeakers. The effect is the same with sound as it is for light.

A key element in a double-slit experiment is using waves that have a single wavelength. If more than one wavelength is present, a more complex and less symmetrical interference pattern would occur. For sound waves, the experiment can be conducted using two speakers both playing a tone with the same frequency. For light waves, the experiment must be conducted using monochromatic light, which is light with a single constant wavelength and a specific color.

Polarization of Light

Light waves are electromagnetic waves consisting of both an oscillating electric field and an oscillating magnetic field. These two oscillations sustain each other and allow light to propagate independently, even through a vacuum. The electric-field and magnetic-field waves are both transverse waves where the oscillating fields are perpendicular to the direction of wave travel. In addition, the electric-field oscillation is perpendicular to the magnetic-field oscillation. Working with both of these perpendicular oscillations is complex. As a result, light waves are often simplified as a single transverse wave involving only the electric-field oscillation.

The electric field is capable of doing work on charges. As a result, the polarization of electromagnetic waves is viewed from the perspective of the electric field's transverse wave form. The poles of the oscillation are the crests and troughs of the electric-field wave, which maintain their orientation as light waves propagate. The plane in which the electric field oscillates is known as the **plane of polarization**. In Figure 17.6, the electric field is oscillating in the x-y plane, and this electromagnetic wave is polarized along the y-axis.

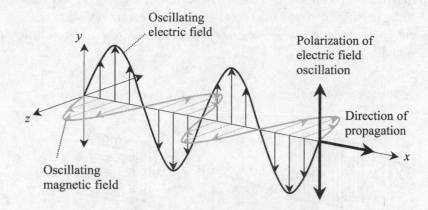

FIGURE 17.6 Electromagnetic wave

Light waves can be polarized in any direction. Light arriving from the Sun consists of countless waves, each polarized in random directions. Light from the Sun, as well as light from other conventional light sources, is unpolarized. The diagrams in Figure 17.7 are simplified by showing the polarization of light for two light sources coming directly out of the page. Figure 17.7(a) represents a vertically polarized electric-field oscillation. Figure 17.7(b) consists of electric-field waves polarized in a variety of directions and is an example of unpolarized light.

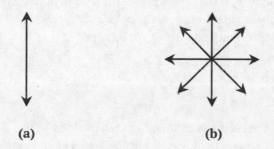

(a)　　　　　　　　**(b)**

FIGURE 17.7 Polarized (a) and unpolarized (b) light

If unpolarized light passes through a **polarizing filter**, it will become polarized. A polarizing filter contains a transparent sheet imbedded with long, organic molecules oriented in only one direction. It is similar to the bars of a jail cell, where some things can pass through and others cannot. Only the light rays oscillating in one direction will pass through the polarizing filter. Light oscillating in all other directions is blocked.

When two polarizing filters have the same orientation, light is polarized in the same manner as if only one filter is present. When two polarizing filters are perpendicular, they block all waves in any orientation. No light passes through. If the filters start in the position shown in Figure 17.8(a) and one filter is turned about the axis of propagation, the light passing through will become dimmer and dimmer. When the filter has turned 90 degrees, all light will be blocked. Polarizing filters are used in 3-D glasses to view movies. A 3-D movie is actually two polarized movies superimposed on the screen. One lens of the 3-D glass allows vertically polarized light to pass through, and the other lens allows horizontally polarized light to pass through. Each eye is watching a different movie. The brain interprets the resulting images as three dimensional.

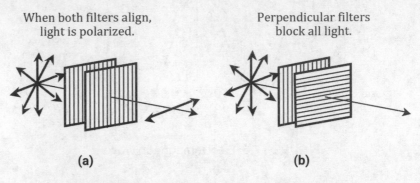

When both filters align, light is polarized.

Perpendicular filters block all light.

(a)   (b)

FIGURE 17.8 Effect of two polarizing filters

When light strikes surfaces at certain angles, it can result in some of the reflected light's becoming polarized. The direction of the polarized waves is parallel to the reflecting surface. This intensifies the brightness of the light entering the eye and is often referred to as "glare." This can be seen when the sun is low in the sky and the sunlight's reflecting off the horizontal surface of water or the hood of a car makes it difficult to look in the direction of the setting or rising sun. Polarizing sunglasses are actually polarizing filters that block the intensified horizontally polarized light seen under these conditions.

Color

Visible light consists of the colors extending from red to violet in the electromagnetic spectrum. To see specific colors, light waves must be aimed at and enter the eye in order to stimulate the photoreceptors in the retina.

Absorption and Reflection

When light strikes a surface, some wavelengths will be absorbed and others will reflect. If light is absorbed, it cannot be seen. Only the reflected rays bouncing off of an object are capable of entering the human eye. When objects are observed with the eye, the reflected colors are seen. For example, think of green leaves. Leaves reflect the wavelengths of light that appear green. This means that other wavelengths of light are being absorbed.

Dispersion

White light arriving from the Sun is composed of all the wavelengths of visible light. These wavelengths can be separated into distinct colors of light through a process known as **dispersion**. This process is commonly demonstrated using a prism. Dispersion takes advantage of both refraction and geometric optics. When white light strikes a prism at an angle, the light separates (disperses) into individual colors. Each color of light has a slightly different wavelength and index of refraction when it moves through the prism. As a result, each color bends at a slightly different angle as it enters and leaves the prism, as shown in Figure 17.9.

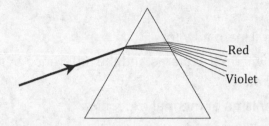

FIGURE 17.9 Dispersion of light through a triangular prism

Dispersion in water droplets together with refraction and reflection are responsible for the appearance of rainbows.

Scattering

When light rays in the atmosphere strike particles in the air, the light rays are reflected in various directions. This process is known as **scattering**. Shorter wavelengths of light are scattered the most. If atmospheric particles are large compared to the wavelength of light, such as water droplets in clouds, then scattering is simply due to the reflection of sunlight. Since sunlight includes all colors in the visible spectrum, clouds appear white on sunny days. If atmospheric particles are small compared to the wavelength of light, such as the transparent molecules that make up air, then scattering is due to the interaction between the light waves and electrons in the atoms of these molecules. This type of scattering is dependent on the wavelength of light. All wavelengths of light in the visible spectrum are scattered. However, shorter-wavelength blue light is scattered the most. As a result, more blue light reaches observers on Earth, giving the sky an overall blue appearance.

CHAPTER 18

Thermal Properties

Learning Objectives

In this chapter, you will learn how to:

o Define and explain thermal systems and the difference between thermal energy and temperature

o Explain the mechanisms of thermal expansion

o Identify the properties of ideal gases

o Define heat and describe heat transfer

o Analyze the heating and cooling of solids, liquids, and gases

Thermal systems consist of all the atoms that make up solids, liquids, and gases. These atoms are in constant random motion. That motion affects the temperature and thermal energy of all of the atoms grouped together as a larger system. Temperature differences between systems will cause energy to flow as heat from the hotter system to the colder system. Temperature, thermal energy, and heat are affected by the physical properties of the atoms making up each system.

Table 18.1 lists the variables discussed.

Thermal Systems

In mechanics, the focus was often on the motion of a single object, such as a block sliding down an incline. However, when several objects experience the same motion, such as two blocks tied together by a string, then they can be treated as a single system. This allows the problem to be solved as though the blocks and the string are one larger object. Treating several objects experiencing the same conditions as one aggregate object is extremely useful in thermal physics.

The objects involved in thermal-physics problems are the countless atoms and molecules that make up substances. Working with single atoms is not possible. As a result, quantities and equations that describe the particles acting together as a single system have been established. The system might be an ice cube that melts into a liquid and vaporizes into a gas. It could be an iron rod that is heated and expands. It could even be a gas trapped in a cylinder with a movable piston. Some aspects of thermal physics involve analyzing the individual objects (atoms) within the larger system. Other problems will be concerned with the properties, trends, and quantities that describe the system as a whole.

TABLE 18.1 Variables That Describe Thermal Properties

New Variables	Units
T = temperature	kelvin
E_{th} = thermal energy	J (joules)
M = molar mass	kg/mol (kilograms per mole)
α = coefficient of linear expansion	1/K or K^{-1} or 1/°C or $°C^{-1}$
L = length	m (meters)
P = pressure	Pa (pascals)
V = volume	m^3 (meters cubed)
n = number of moles	mol (moles)
R = universal gas constant	8.31 J/mol · K (Joule per Kelvin times mole)
Q = heat	J (joules)
k = thermal conductivity	W/m · K (watts per meter · kelvin)
c = specific heat capacity	J/kg · K (joules per kilogram · kelvin)
L = latent heat	J/kg (joules per kilogram)

Thermal Energy

The atoms composing solids, liquids, and gases are in constant motion. Thus, each atom has a tiny amount of kinetic energy. The chemical bonds and intermolecular forces holding substances together store tiny amounts of potential energy. **Thermal energy** is the sum of the microscopic kinetic and potential energies of all of the atoms making up the system under investigation.

Thermal energy increases when substances become hotter. When solids become hotter, their molecules vibrate faster and increase in kinetic energy. When liquids and gases become hotter, the randomly moving molecules move faster and increase in kinetic energy.

Converting Mechanical Energy into Thermal Energy

Thermal energy and mechanical energy are both measured in joules. According to the law of conservation of energy, these two forms of energy are capable of transforming from one type into the other as long as the total amount of energy in the system is conserved. We will explore instances where mechanical energy is transformed into thermal energy. This transfer occurs in two common ways: friction and inelastic collisions.

When a block slides down a rough incline, the rubbing of the block against the incline makes the molecules in both the block and the incline vibrate faster. Their thermal energies increase. Conservation of energy dictates that the thermal energy gained must equal the kinetic energy lost by the slowing block. The kinetic energy lost by the block is an energy change, and an energy change is known as work. In this case, the work of friction transfers kinetic energy from the block into thermal energy. The work of friction is equal to the kinetic energy lost by the block and is also equal to the thermal energy generated.

In inelastic collisions, the colliding objects impact one another. The impact causes the atoms within the objects to vibrate faster. In an inelastic collision, kinetic energy is lost. As with friction, the thermal energy generated in the inelastic collision is equal to the kinetic energy lost during the collision.

Temperature

Temperature is a relative measure of hot and cold. The **Celsius** and Fahrenheit temperature scales were established to encompass common extremes of hot and cold experienced by humans. The extremes of freezing and boiling water were used to set the 0 degree and 100 degree marks on the Celsius scale. Once established and accepted, the Celsius temperature scale became a means to quantify temperatures so that hot and cold objects could be numerically compared with one another.

Temperature reflects the speed and microscopic kinetic energy of the molecules of a substance. This means that the temperature of an object reflects the thermal energy of the object. Raising the temperature of an object increases its hotness. This causes the atoms of the substance to move faster, increasing the thermal energy. In order to relate temperature directly to thermal energy mathematically, a less arbitrary temperature scale is needed. Experimentation with gases resulted in projecting a temperature at which all molecular motion should cease. At this temperature, gas molecules would have no microscopic kinetic energy and therefore no thermal energy. The resulting temperature scale is the **Kelvin** temperature scale. Absolute zero, 0 kelvin, on this scale is equal to $-273°C$. It is the point where both temperature and thermal energy equal zero. As a result, formulas could be developed that use the Kelvin temperature to calculate thermal energy.

Which scale should be used for calculations? If a formula contains a T for temperature, then the Kelvin temperature must be used. However, if a formula contains a ΔT, then either the Celsius or Kelvin scale can be used. Why? Although the Celsius and Kelvin temperature scales have different zero points, 1 degree of change on the Celsius temperature scale is equal to 1 degree of change on the Kelvin temperature scale. Therefore, a change in temperature, ΔT, will be the same using either scale. When in doubt, it is always safer to use the Kelvin scale. To convert from degrees Celsius to kelvin, add 273.

Thermal Expansion

When a solid or a liquid is heated, the atoms and molecules vibrate faster, causing a substance that is heated to expand. This process is known as **thermal expansion**. Conversely, a substance that is cooled will contract. Heating and cooling will change the length of linear objects, the surface area of two-dimensional objects, and the volume of three-dimensional objects.

In conceptual problems, any type of object can be given and the effects of heating or cooling it may be asked. Essentially, the entire object expands or contracts proportionally.

Linear Expansion

When a linear object, such as a metal rod, is heated, its length will increase, as shown in Figure 18.1.

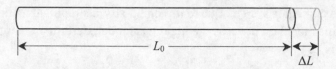

FIGURE 18.1 Linear expansion

The amount that the rod increases in length, ΔL, is calculated as follows:

$$\Delta L = \alpha L_0 \Delta T$$

$$\Delta L = \alpha L_0 (T_f - T_i)$$

The change in length, ΔL, is equal to the product of the coefficient of linear expansion, α, the original length, L_0, and the change in temperature, ΔT. The coefficient of linear expansion is a physical property of the substance that is expanding. Its value depends on the composition of the object. When a substance is heated, its final temperature, T_f, will be larger than its initial temperature, T_i. This heating will cause ΔL to become positive, indicating that length is increasing. However, when a substance is cooled, T_i will be larger than T_f. This cooling will result in a negative ΔL, indicating that length is decreasing. To find the new length, L, simply add the change in length, ΔL, to the starting length, L_0.

The linear-expansion formula can be used to solve for the expansion of a rectangular area. A rectangle has a length and a width. Simply solve for linear expansion twice, once for each dimension. Then, multiply the new length by the new width to find the new area.

EXAMPLE 18.1

Linear Expansion

A 5.0-meter-long iron rod is heated from 0°C to 100°C. The linear coefficient of expansion for iron is 12×10^{-6} K^{-1}. Determine the change in length of the rod.

WHAT'S THE TRICK?

Use the linear-expansion formula, but read the problem carefully. Is the problem looking for the change in length or the new length after expanding or contracting? This problem simply requests the change in length.

$$\Delta L = \alpha L_0 \Delta T$$
$$\Delta L = \alpha L_0 (T_f - T_i)$$
$$\Delta L = (12 \times 10^{-6}°C^{-1})(5.0 \text{ m})(100°C - 0°C)$$
$$\Delta L = 6.0 \times 10^{-3} \text{ m}$$

Ideal Gases

Gases are extremely chaotic. They consist of countless particles moving at different speeds in every possible direction. These particles collide with each other and with surfaces. Analyzing the behavior of a gas by examining individual gas molecules is impossible. Instead, a gas is treated as a single aggregate system. Due to its complexity, a simplified model of a gas system was developed. This model is known as the **ideal gas** model. Certain assumptions are made about ideal gases:

- The particles are so small that the volume of the particles is negligible.
- The attraction between particles is zero (no microscopic potential energy).
- The particles are in constant random motion (microscopic kinetic energy).
- Collisions between particles and with surfaces are perfectly elastic.

These assumptions make it possible to analyze ideal gases using the mathematical relationships discussed in the sections that follow. The behavior of a real gas at normal temperatures is nearly identical to that of an ideal gas.

Since the particles of a gas move at varying speeds in all directions, calculations involving specific directional velocities are impossible. As a result, only the average speed of the gas particles can be determined. As temperature increases, average speed increases. The average speed, v_{rms}, is related to the Kelvin temperature, T, of the gas according to the following formula.

$$v_{rms} = \sqrt{\frac{3RT}{M}}$$

The speed of a gas is changed by the square root of the factor applied to the Kelvin temperature. If the Kelvin temperature is doubled, the average speed of the gas particles increases by $\sqrt{2}$. The variable R is the universal gas constant. In physics, it has the value of 8.31 joules per mole • kelvin. The variable M is the molar mass of the gas particles in kilograms per mole.

The energy of a gas is related to the average speed and kinetic energy of the gas particles. Since the attraction between the particles of an ideal gas is assumed to be zero, the only type of energy that gas particles possess is the microscopic kinetic energy due mainly to their speed. Since all the particles have different speeds, the kinetic energy, K, of the gas particles is also an average. It is calculated as follows.

$$K_{avg} = \frac{3}{2} k_B T$$

The variable k_B is Boltzmann's constant, $k_B = 1.38 \times 10^{-23}$ joules/kelvin. The formula is valuable in demonstrating that the average kinetic energy of a gas is directly proportional to its Kelvin temperature. If the Kelvin temperature of a gas is doubled, the average kinetic energy of the gas particles also doubles.

The trends relating temperature, gas-particle speed, and the energy of the gas may be the most important information to retain. As temperature increases, molecules move faster and their kinetic energy increases. If the attraction between the gas particles is assumed to be zero, the microscopic potential energy between gas particles is also zero. This means that the thermal energy of a gas is essentially the sum of the kinetic energies of all of the gas particles.

Pressure

When a particle of gas collides with a surface, such as the walls of a container, the momentum change (impulse) experienced by the gas particle causes a force to be applied to the surface. Calculating the force of each particle of gas and adding them up would be an impossible task. However, the aggregate force of all gas particles treated as a single system striking a specific area can be measured and is a valuable quantity. The force, F, due to all the collisions of gas particles striking 1 square meter of surface area, A, is known as **pressure**, P.

$$P = \frac{F}{A}$$

Pressure is measured in pascals (Pa). A pascal is equal to a newton per meter squared (N/m^2).

Once again, the overall trends are extremely important. As the temperature of a gas increases, the particles of the gas move faster and increase their speed. The kinetic energy of the gas particles increases as does the thermal energy of the gas as a whole. In addition, the faster and more energetic gas particles experience greater momentum changes when they strike surfaces, resulting in an increase in the pressure of the gas.

Ideal Gas Law

The ideal gas law is a mathematical relationship relating the pressure, P, volume, V, number of moles, n, and temperature, T, of an ideal gas.

$$PV = nRT$$

Volume in physics is measured in meters cubed, m^3. The variable n represents the number of moles of gas particles. A mole is similar to terms such as "pair" or "dozen." These terms bring specific numbers to mind (pair = 2 and dozen = 12). A mole is a specific number of particles. One mole is 6.02×10^{23} particles, and this value is known as Avogadro's number. Avogadro's number is not needed for the ideal gas law; just the number of moles is needed. If a problem specifies one mole (1 mol) of gas, then $n = 1$. For two moles of gas, $n = 2$, etc. The ideal gas constant, R, is equal to 8.31 joules per mol • kelvin.

Relationships between key variables in the ideal gas law, $PV = nRT$:

1. **Pressure and volume are inversely proportional to each other.** For example, if the temperature and the moles of gas are held constant, then a decrease in volume is compensated for by an increase in pressure.

2. **Pressure is directly proportional to the Kelvin temperature of a gas.** For example, increasing the temperature while holding the volume and the number of moles constant will increase the pressure of a gas.

3. **Volume is directly proportional to the Kelvin temperature of a gas.** For example, increasing the temperature while holding the pressure and the number of moles constant will increase the volume of a gas.

EXAMPLE 18.2

Ideal Gas Law

A gas is trapped in a cylinder with a movable piston. How is the pressure of the gas affected if the temperature of the gas doubles while the piston moves inward, reducing the volume by half?

WHAT'S THE TRICK?

The ideal gas law is the key. The problem states that the gas is trapped, implying that the number of moles remains constant. Any value remaining constant, such as the number of moles and the gas constant, cannot cause change. Ignore them.

$$PV = nRT$$

$$(4P)\left(\frac{1}{2}V\right) = nR(2T)$$

The pressure must quadruple in order to maintain the equality.

Heat and Heat Transfer

Heat is often misinterpreted by beginning physics students since the word *heat* is used incorrectly in everyday life. When students touch a warm object, they think it contains a large quantity of heat energy. This is not necessarily correct. The temperature felt by touching a warm object is a result of the thermal energy of the vibrating particles in that object. Try not to confuse the word *heat* with the word *hot*. **Hot** refers to the temperature of an object alone. Temperature is a measure of the average kinetic energy of the particles in that object. **Heat** refers not only to temperature but also to the number of particles (mass) that are involved. Consider a hot cup of coffee and the ocean. The coffee is hotter in temperature. However, the ocean contains more heat since it contains significantly more particles than the coffee. Then what is heat? Heat is similar to work. Work is a mechanical change in energy that can be seen by the eye (macroscopic). When an object is pushed by a force through a distance, its kinetic energy changes. Work equals the change in macroscopic kinetic energy. **Heat**, Q, is a change in thermal energy that cannot be seen by the eye (microscopic). When a flame is applied to an object, the object's thermal energy changes. Heat equals the change in thermal energy. While work is the quantity of mechanical energy transferred from one system to another, heat is the quantity of thermal energy transferred from one system to another system.

Heat Transfer

Heat transfer is the process of transferring thermal energy from one system to another. In order for heat transfer to take place between two systems, the systems must be at different temperatures. The natural direction of heat transfer is from the high-temperature system to the low-temperature system. The particles in the high-temperature system are vibrating or moving faster. When they come into contact with the slower particles in the low-temperature system, collisions transfer energy (heat) from the high-temperature system to the low-temperature system. The particles in the high-temperature system lose energy, slow down, and become cooler. The particles in the low-temperature system gain energy, speed up, and become hotter. Energy transfer continues until both systems reach the same temperature. **Thermal equilibrium** describes the condition when two objects have the same temperature, and no net heat transfer will take place between them on their own. Consider the coffee and ocean example. Pouring the coffee into the ocean transfers heat from the coffee, which is hotter, to the ocean, which is cooler. Both will ultimately achieve the same temperature, or thermal equilibrium. However, the direction of heat transfer always goes from the object with the higher temperature to that with the lower temperature.

Heat transfer occurs by three methods. In **conduction**, heat is transferred when two objects at different temperatures physically touch each other. **Convection** is heat transfer by fluids (liquids and gases). **Radiation** is heat transfer due to the absorption of light energy.

Conduction

Transfer of heat by **conduction** requires objects to touch each other physically. This normally involves a hotter solid object's touching a colder solid object. This is similar to the process of conduction in electricity, where charges transfer energy from one conductor to another conductor when they touch each other. In fact, substances that are good conductors of electricity, such as metals, are also good conductors of heat.

Rate of Heat Transfer and Thermal Conductivity

The **rate of heat transfer**, $Q/\Delta t$, through an object is dependent on the length, L, the heat must travel through the object; the cross-sectional area of the object, A; the temperature difference between the ends of length L; and the thermal conductivity, k, of the object. The following formula solves for the rate of heat transfer.

$$\frac{Q}{\Delta t} = \frac{kA\Delta T}{L}$$

Thermal conductivity is a physical property of an object. It indicates how well heat is conducted through an object. Every substance has a unique thermal conductivity. The higher the thermal conductivity of the substance, the faster heat transfers through it. Copper has a thermal conductivity of 400 watts/meter • kelvin, while stainless steel has a thermal conductivity of 14 watts/meter • kelvin. This means that copper pots and pans transfer heat through them at a much faster rate, allowing for faster meal preparation and quicker adjustments in temperature when cooking.

Insulators can be used to reduce the rate of heat transfer. This is similar to the process of insulation in electricity. In fact, substances that are good insulators of electricity, such as nonmetals, are also good insulators of heat. Insulation is used in homes to prevent summer heat from entering and winter heat from escaping. Insulating materials need to prevent the rapid transfer of heat, and these substances will have very low thermal conductivities.

Thermal conductivity also explains why objects at the same temperature can feel like they have different temperatures when touched. Metals have high thermal conductivities, and wood has a very low thermal conductivity. Even if a piece of metal and a piece of wood are both at room temperature, the metal will feel cooler to the touch. The human body is warmer than room temperature. When these objects are touched, heat transfers from the fingers into both the metal and the wood. However, heat transfers into the metal at a faster rate. The rapid loss of body heat as it moves into the piece of metal makes the fingers feel colder.

Convection

Heat transfer through **convection** involves fluids. Fluids can flow. This means both liquid and gases are fluids. When a hot fluid mixes with a cold fluid, thermal energy is transferred as the faster-moving particles collide with the slower-moving particles. Heating a home is an example of convection. Hot air flows through ducts in the home, eventually pouring into a room, much like a river pours into a lake. A moving fluid with a different temperature than its surroundings is known as a **convection current**.

Radiation

Heat can also be transferred by electromagnetic **radiation**. Light waves carry energy that can be absorbed when the waves strike objects. The absorption of light will increase the thermal energy of an object, causing its temperature to increase.

Heating and Cooling

Specific Heat

If a substance is heated, it will absorb energy, and its temperature will increase. If a substance is cooled, it will lose energy, and its temperature will decrease. Every substance absorbs or loses energy at a set rate, which can be quantified. **Specific heat**, c, is the amount of heat needed to raise the temperature of 1 kilogram of a substance by 1 kelvin. For example, the specific heat of liquid water is 4,190 joules per kilogram • kelvin. In order to raise the temperature of 1 kilogram of water by 1 kelvin, 4,190 joules of heat must be transferred to the water. The following equation is used to solve for the heat needed to change the temperature of a substance with a mass of m and a temperature difference of ΔT.

$$Q = mc\,\Delta T$$

The specific heat, c, is a physical property that is unique for every substance. In addition, each state of matter has a unique specific heat. For example, the specific heat of solid water, c_s, is 2,090 joules per kilogram • kelvin. However, the specific heat of liquid water, c_L, is 4,190 joules per kilogram • kelvin.

Liquid water has the highest specific heat of substances commonly found on Earth. As a result, a tremendous amount of energy is needed to increase the temperature of water. Similarly, water can retain a tremendous amount of energy. This allows for the temperature of water to change quite slowly in either direction. It also allows water to be a **heat sink**, whereby it can store large quantities of energy in places such as rivers, lakes, and particularly oceans.

EXAMPLE 18.3

Specific Heat

The specific heat of aluminum is 900 joules/kilogram · kelvin. How much heat is required to raise 2.0 kilograms of aluminum 10°C?

WHAT'S THE TRICK?

For quantities other than 1 kilogram and 1 kelvin, use the equation

$$Q = mc\,\Delta T$$

$$Q = (2.0 \text{ kg})(900 \text{ J/kg} \cdot \text{K})(10 \text{ K})$$

Note that a change of 10°C is the same as a change of 10 K.

$$Q = 18,000 \text{ J}$$

Phase Changes

Matter on Earth is commonly found in three phases: solid, liquid, or gas. A **phase change** is when a substance physically changes from one phase to another. Phase changes occur when substances reach critical temperatures. Solids change into liquids at their melting point, and liquids change into gases at their boiling point. The melting and boiling points are two important physical properties of a substance. Different substances will have different melting and boiling points. However, every substance has a set melting and boiling point under specific environmental conditions.

Heat of Transformation (Latent Heat)

When a substance is heated, its temperature will increase until it reaches the critical temperature (melting point or boiling point) at which a phase change can occur. At this critical temperature, all the thermal energy added to the substance is used to conduct the phase change. The **heat of transformation**, also known as **latent heat**, is the energy added during the phase change. This energy is used to weaken the intermolecular forces that hold molecules together as solids and liquids. Since all the energy added is involved in the transformation, the temperature of the substance does not change during the phase change. When all of the substance has completed the phase change, the temperature of the substance can then resume its rise.

The heat of transformation, or latent heat, L, is a physical property of a substance. Each substance has a unique value for this quantity. The **heat of fusion** or **latent heat of fusion**, L_f, is the heat energy needed to convert 1 kilogram of a substance from its solid form to its liquid form. The **heat of vaporiztion** or **latent heat of vaporization**, L_v, is the heat energy needed to convert 1 kilogram of a substance from its liquid form to its gaseous form. The heat of vaporization is always significantly larger than the heat of fusion. More energy is required for the phase change from liquid to gas. The values given for latent heat will be the amount of heat needed for exactly 1 kilogram of a substance. To solve for the heat needed in phase changes involving a substance with mass m, use the following formula.

$$Q = mL$$

EXAMPLE 18.4

Heat of Transformation

Water has a heat of fusion of 3.33×10^5 joules per kilogram and a heat of vaporization of 22.6×10^5 joules per kilogram. How much heat energy is needed to melt 2.0 kilograms of ice at 0°C?

WHAT'S THE TRICK?

Melting involves the heat of fusion. The given heat of vaporization is a distracter.

$$Q = mL$$

$$Q = (2.0 \text{ kg})(3.33 \times 10^5 \text{ J/kg}) = 6.66 \times 10^5 \text{ J}$$

CHAPTER 19

Thermodynamics

Learning Objectives

In this chapter, you will learn how to:

o Define internal energy and examine its role in thermodynamics

o Summarize how work and heat affect internal energy

o Apply the first law of thermodynamics to solve problems

o Explain the nature of entropy

o Analyze the consequences of entropy and the second law of thermodynamics

Thermodynamics is the branch of physics dealing with thermal energy and heat. In beginning physics classes, thermodynamics focuses on a system consisting of gas particles that are contained in a cylinder with a movable piston. Working with countless particles in random motion is an impossible task. Therefore, thermodynamics examines gases as a single system.

Table 19.1 lists the variables used in thermodynamics.

TABLE 19.1 Variables Used in Thermodynamics

New Variables	Units
U — internal energy	J (joules)
e = efficiency	% (percent)

Internal Energy

Internal energy, U, is the total energy of a system. Adding up all possible energies that compose the total energy of a system would be extremely complex. However, doing this is not necessary. Thermodynamics is not concerned with the actual value of internal energy, U, but rather with the change in internal energy, ΔU. It is the change in internal energy that makes engines operate. In thermodynamics, all but one of the energies making up internal energy are held constant. Thermal energy, E_{th} or $E_{thermal}$, is the only energy that is changing. The change in all other energies will be equal to zero.

$$\Delta U = \Delta E_{thermal}$$

The thermal energy of a gas is the sum of all the microscopic kinetic energies of the randomly moving gas particles. This energy can be calculated and results in the following equation for the change in internal energy of a gas.

$$\Delta U = \frac{3}{2} nR \, \Delta T$$

This equation has been included here to show the important direct relationship between temperature change and the change in internal energy. In addition, the sign on the change in internal energy in this equation is the key to setting the correct sign on other thermodynamic variables.

During a thermodynamic process there are three possibilities for the value of internal energy.

1. **Internal energy increases.** If energy is added to the system, the internal energy increases, $+\Delta U$. This is accompanied by an increase in the temperature of the system, $+\Delta T$.

2. **Internal energy decreases.** If energy is removed from the system, the internal energy decreases, $-\Delta U$. This is accompanied by a decrease in the temperature of the system, $-\Delta T$.

3. **Isothermal process.** This is a thermodynamic process where the temperature remains constant, $\Delta T = 0$. If the temperature is not changing, then the internal energy is not changing, $\Delta U = 0$.

Energy Transfer in Thermodynamics

To visualize a thermodynamic system, think of a gas trapped in a cylinder with a movable piston. Only the gas particles constitute the system. The cylinder and piston are part of the environment. They simply contain the gas in an adjustable and measurable volume. This system is shown in Figure 19.1.

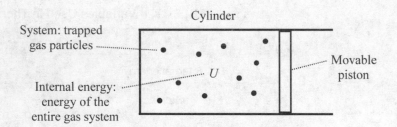

FIGURE 19.1 A thermodynamic system

Work and heat are the two processes that can transfer energy into and out of the gas system shown in Figure 19.1. Work and heat are in units of joules. Work and heat are the changes to state functions, such as internal energy U, kinetic energy K, and the energy of a spring U_s.

Work

Work, W, is a macroscopic (visible) transfer of energy involving a net force, ΣF, acting through a distance, d. Work is a mechanical way to add or remove energy from the gas system by physically moving the piston. The force that moves the piston is created by pressure acting on the inside and outside areas of the piston. The pressure of the gas inside the cylinder pushes the piston outward. Forces from the environment, such as the pressure of the atmosphere surrounding the cylinder, push the piston inward.

Moving the piston definitely involves a volume change, ΔV, but what about pressure? Under some conditions, the pressure of the gas changes when work moves the piston. Under special conditions, the pressure remains constant as the piston moves. This is known as an **isobaric** process. Under these special conditions, work is a product of the constant pressure and changing volume.

$$W = -P\Delta V$$

Whether the pressure of the gas is changing or is constant, there are three possible outcomes. Which particular outcome occurs depends on how the magnitudes of the pressures inside and outside the cylinder compare to one another.

1. **Work done on the gas.** If the pressure outside the cylinder is greater than the pressure inside the cylinder, then work is done *on the gas* while moving the piston inward. The volume of the gas is compressed, $-\Delta V$. The environment adds energy to the system, increasing the internal energy, $+\Delta U$. Work is a change in energy. The work done *on the gas* is positive, $+W$, increasing the internal energy of the system.

2. **Work done by the gas.** If the pressure inside the cylinder is greater than the pressure outside the cylinder, then work is done *by the gas* while moving the piston outward. The volume of the gas expands, $+\Delta V$. In the process, the gas uses its energy to push the piston outward, which causes internal energy to decrease, $-\Delta U$. Therefore, the work done *by the gas* is negative, $-W$.

3. **Isometric (isochoric) process.** This is a thermodynamic process where the volume of the gas remains constant. This occurs when the pressure of the gas system inside the cylinder is the same as the pressure of the environment outside the cylinder. The net force is zero. The piston cannot move, $\Delta V = 0$, and *no work is done*, $W = 0$.

Heat

Heat (Q) is a microscopic (invisible) transfer of thermal energy between objects having different temperatures. Heat can be added or removed by touching a **heat reservoir** to the cylinder. An example of a heat reservoir could be air or water surrounding the cylinder. The energy transferred into or out of the reservoir is negligible compared with the huge size of the reservoir. As a result, heat reservoirs have constant temperature and transfer heat at a constant rate. A hot reservoir has a temperature greater than the gas in the cylinder. A cold reservoir has a temperature lower than the gas in the cylinder.

Heat flows from high temperature to low temperature. There are three possible outcomes for heat transfer depending on the temperature of the heat reservoir as compared with the temperature of the trapped gas in the cylinder.

1. **Heat added.** If the heat reservoir has a higher temperature (hot reservoir) than the gas inside the cylinder, heat flows from the reservoir *into the cylinder*. Internal energy increases, $+\Delta U$. Therefore, heat added is considered positive, $+Q$.

2. **Heat removed.** If the heat reservoir has a lower temperature (cold reservoir) than the trapped gas inside the cylinder, heat flows *out of the cylinder* into the reservoir. Internal energy decreases, $-\Delta U$. Therefore, heat removed is considered negative, $-Q$.

3. **Adiabatic process.** This is a thermodynamic process where *no heat is added or removed*. This occurs when the heat reservoir and the trapped gas are in thermal equilibrium (same temperature). Heat transfer requires a temperature difference. So, no net heat flow occurs between objects in thermal equilibrium, $Q = 0$.

Energy Model Summarized

Figure 19.2 summarizes the interactions among internal energy, work, and heat. The signs on each variable are linked to their effect on the internal energy, ΔU, of the system (the gas trapped in the cylinder). Table 19.2 summarizes additional information and trends.

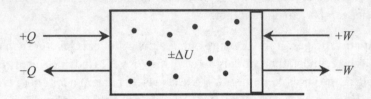

FIGURE 19.2 Thermodynamics energy model

TABLE 19.2 Interactions Among Internal Energy, Work, and Heat

If You See . . .	Result	Related Events	Result
Increase in internal energy	$+\Delta U$	Increase in temperature	$+\Delta T$
Decrease in internal energy	$-\Delta U$	Decrease in temperature	$-\Delta T$
Constant internal energy	$\Delta U = 0$	Isothermal: constant temperature	$\Delta T = 0$
Work done on the gas	$+W$	Volume of gas compresses	$-\Delta V$
Work done by the gas	$-W$	Volume of gas expands	$+\Delta V$
No work is done	$W = 0$	Isometric: constant volume	$\Delta V = 0$
Heat added	$+Q$	Heat reservoir is hotter	
Heat removed	$-Q$	Heat reservoir is colder	
No heat added or removed	$Q = 0$	Adiabatic: no heat transfer	

First Law of Thermodynamics

The **first law of thermodynamics** is a statement of conservation of energy for thermal processes. Work, W, and heat, Q, are the only two processes that can affect the internal energy, U, of a gas. Their effect can be summarized as follows.

- For a system where internal energy comprises only thermal energy, the change in the internal energy of the system is equal to the energy transferred into or out of the system by work and heat.

- The previous statement can be summarized as an equation.

$$\Delta U = Q + W$$

Entropy

The natural flow of heat is from high temperature to low temperature. Systems with different temperatures contain particles that are moving at different average speeds. When these systems mix, all types of collisions occur. However, mathematical probability drives the direction of heat transfer. The warmer system has more fast-moving particles, while the cooler system has more slow-moving particles. Therefore, collisions between fast particles and slow particles are more likely. In these collisions, energy is transferred from the fast particles to the slow particles. Energy continues to transfer until both systems have the same average particle speeds. Therefore, the equilibrium state is the statistically most probable state that can occur.

Entropy is a way to quantify the probability of finding a system in a particular state. Figure 19.3(a) shows two gases separated by a movable wall. The gas in the left compartment initially has more molecules than the gas in the right compartment. Figures 19.3(b) and 19.3(c) show two possible states that the system can be found in after the wall has been removed.

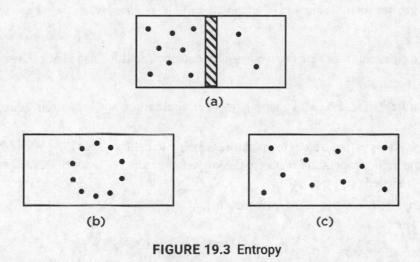

FIGURE 19.3 Entropy

Figures 19.3(b) and 19.3(c) are snapshots of the system at an instant in time. The gas particles are in random motion, and both of the resulting diagrams are actually possible. However, the highly organized pattern in Figure 19.3(b) is about as likely as winning millions in a lottery. It has a low probability, and therefore it has low entropy. Random states like the one seen in Figure 19.3(c) are statistically more likely to occur. They are more probable and thus have higher entropy. Probability drives systems toward random and disordered equilibrium states with higher entropy. Therefore, entropy is associated with messiness or randomness. The more random a system is, the greater its entropy. In addition, the natural trend is toward equilibrium and toward greater entropy.

Look again at Figure 19.3(c), the most probable result when the wall was removed. What is the likelihood that the molecules in Figure 19.3(c) will spontaneously separate back into their original compartments, returning to their locations in Figure 19.3(a), and then remain in those positions without a wall to hold them? This scenario is not likely and does not take place. This means that removing the wall is an irreversible step. Entropy drives thermodynamic processes toward equilibrium. It ensures that isolated (no environmental interference) thermal processes are irreversible.

Second Law of Thermodynamics

The **second law of thermodynamics** addresses entropy, the drive toward equilibrium and its irreversible nature.

- The entropy of an isolated system cannot decrease.
- The entropy of isolated systems always increases until the system reaches equilibrium.
- Once at equilibrium, the entropy of the system remains constant.

An **isolated system** is a system that follows natural tendencies and does not interact with the surrounding environment. When systems are not isolated, natural tendencies may be reversed as long as energy is supplied to the system by the environment. For example, the natural tendency is for a waterfall to flow downward. This can be reversed. As the Sun's radiant energy is added to the water, the water rises through evaporation to continue the water cycle. The environment must expend a great deal of energy in an effort to interfere with the natural tendency of the system. The energy required to carry the water to the top of the waterfall is greater than the amount of energy that will be released when the water falls on its own.

The following general trends are consequences of entropy and the second law of thermodynamics.

- The natural tendency is for systems to move to equilibrium and for entropy (disorder) to increase.
- When systems with different temperatures come into contact, heat flows spontaneously from the high-temperature region to the low-temperature region until thermal equilibrium is reached.
- Heat engines can never be 100 percent efficient.

CHAPTER 20

Atomic and Quantum Phenomena

In this chapter, you will learn how to:

○ Understand atomic-model history through key experiments and experimenters

○ Define light quantum (photons) and its role in transmitting energy

○ Define the photoelectric effect

The theory that matter is composed of indivisible units, known as atoms, dates back to the ancient Greeks. At the turn of the nineteenth century, a series of experiments led scientists to modify many of their previous ideas about the composition and behavior of atoms. This chapter will discuss several of the experiments leading to these modifications and explain the current understanding of atoms and their quantum behavior.

Table 20.1 lists the variables that will be used.

TABLE 20.1 Variables Used for Atomic and Quantum Phenomena

New Variables	Units
K_{max} = maximum kinetic energy	J (joules)
ϕ = work function	J (joules)
h = Planck's constant	J · s (joule · seconds)

Development of the Atomic Theory

The ancient Greeks first proposed the word *atom* as the name for an indivisible unit of matter. Several observations and prominent experiments have led to a greater understanding of atoms. The following sections include major highlights and scientists involved in the development of the atomic theory.

J. J. Thomson

In 1897, J. J. Thomson discovered that even atoms themselves were divisible when he discovered the electron. Thomson knew that matter had an overall neutral charge, so he theorized that an atom of matter would be a mixture of positive and negative components. His model, often referred to as the "raisin cake model" or "plum-pudding model," visualized the atom as containing positive and negative charges that were distributed throughout the interior of the atom, as shown in Figure 20.1.

FIGURE 20.1 Plum-pudding model of an atom

Ernest Rutherford

Ernest Rutherford, a student of Thomson, began to experiment with what appeared to be charged rays emanating from crystals of uranium. He named these rays alpha and beta. He also determined that they consisted of streams of particles. Rutherford determined that the alpha particle was, in fact, a doubly charged positive ion, which we now know to be a helium nucleus consisting of two protons and two neutrons. The beta ray consists of a negative particle.

To investigate the interior of the atom, Rutherford used a radioactive source to fire alpha particles through a thin sheet of gold foil. A screen sensitive to alpha particles surrounded the gold foil to record the strikes of the alpha particles. If Thomson's model was correct, the even distribution of positive and negative components of gold atoms should not greatly affect the path of the positive alpha particles. As a result, Rutherford expected the fairly heavy alpha particles to pass through the gold atoms with little deflection.

After the experiment was concluded, Rutherford was surprised at the result. Although most of the alpha particles passed through the gold foil as expected, some of the alpha particles were deflected at extreme angles. In fact, a few nearly reversed direction completely. Rutherford likened this to shooting a cannonball at a piece of tissue paper and watching it bounce back. These results prompted Rutherford to propose that the positive region inside the gold atoms (protons were not yet discovered) was actually concentrated in a very tiny nucleus at its center, as illustrated in Figure 20.2.

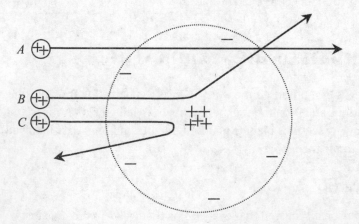

FIGURE 20.2 Rutherford's gold-foil experiment

Most alpha particles, like the one labeled *A* in Figure 20.2, passed through the gold foil without being deflected. A small number of positive alpha particles were deflected at extreme angles, such as particle *B* in Figure 20.2. These particles had to pass close to a dense positive region in order to be deflected in this manner. The most amazing results were those similar to *C* in Figure 20.2. These alpha particles nearly bounced back and must have encountered a very dense positive region head on. Rutherford was able to use the particle traces to map the atom. He proposed that the atom was mostly empty space, allowing the majority of particles through the gold atoms without interference. The deflected traces revealed a very small but extremely positive region at the center of the atom. The results led Rutherford to propose the following characteristics for atoms.

1. Atoms are mostly empty space.
2. The majority of the mass of an atom is concentrated in a tiny central nucleus.
3. The central nucleus is positively charged.
4. The electrons orbit the nucleus in a manner similar to planets orbiting the Sun.

Spectroscopy

Spectroscopy is the study of the light spectrum emitted by luminous objects, such as the Sun, and the light emitted by gas discharge tubes. A gas discharge tube emits light by using a high voltage to shoot electrons through atoms in gaseous form. The electrons collide with the atoms, adding energy to the atoms. Eventually, the excess energy is lost. In the process, light with specific wavelengths is emitted. Each element emits a unique spectrum of light consisting of exact wavelengths. When viewed through a diffraction grating, the colors of the emitted light appear as a series of discrete lines. Each line in the pattern has a single wavelength and color. The patterns formed by these lines differ for each element, and they act like a fingerprint or bar code identifying each element.

Max Planck

In 1900, Max Planck attempted to explain the color spectrum seen when substances were heated to the point of glowing. The light emitted was due to atomic oscillations. The patterns seen could be explained only if Planck assumed that the atomic oscillations had very specific quantities of energy. He suggested that the oscillations were quantized (came in specific quantities). He was able to develop a mathematical relationship for these oscillations. He determined that it was based on a constant, now known as Planck's constant, h.

$$h = 6.63 \times 10^{-34} \text{ joule} \cdot \text{seconds} = 4.14 \times 10^{-15} \text{ electron volt} \cdot \text{seconds}$$

Electron Volts

Planck's constant may be given in units of joule-seconds and/or units of electron volt–seconds. An **electron volt** (eV) is an alternate unit of energy. Atoms are incredibly small, so working with joules of energy is not ideal. It is like measuring the length of a pen with a mile stick. The electron volt is a unit that is scaled to match the size of an atom. The conversion between joules and electron volts involves the same numerical value as the charge on an electron.

$$1 \text{ electron volt} = 1.6 \times 10^{-19} \text{ joules}$$

Albert Einstein

In 1905, Albert Einstein applied Planck's idea of quantization to electromagnetic radiation. He suggested that light is quantized and that it consists of massless, particle-like packets that have a specific quantum (quantity) of energy. These packets of light came to be known as **photons**. Einstein clarified the relationship suggested by Max Planck. Einstein determined that the energy of a photon, E, is the product of frequency, f, and Planck's constant, h.

$$E = hf$$

Wave speed is a function of wavelength and frequency, and the speed of light in a vacuum is c.

$$v = f\lambda$$

$$c = f\lambda$$

The second formula can be rearranged to solve for frequency, $f = c/\lambda$. Then, substitute this into the equation $E = hf$. The result is a useful equation that relates the energy of a photon of light to its wavelength.

$$E = \frac{hc}{\lambda}$$

Niels Bohr

In 1913, Niels Bohr used Einstein's light quanta to suggest a model of the atom that explained why electrons do not fall into the nucleus and why the light emitted from excited atoms produces the observed emission spectra. Bohr theorized that the electrons of atoms could occupy only exact **energy levels**. An energy level is a specific energy state with an exact quantum (quantity) of energy.

The **absorption** and **emission** of specific wavelengths of light are related to the energy difference between these energy levels. When atoms absorb light, the absorbed photons combine with the electrons. The energy of the photons adds to the energy of the electrons. Electrons with this added energy are said to be **excited** and must occupy a higher energy level. When atoms emit light, the excited electrons lose energy by emitting photons. The less energetic electrons must now occupy lower energy levels. The light that is absorbed and emitted is restricted by the energy levels in the atom. Only photons with quanta (quantities) of energy matching the exact difference between energy levels can be absorbed and emitted by an atom. Every atom has unique energy levels, and the difference between energy levels varies from atom to atom. Photons emitted from different elements experience different energy changes, resulting in unique wavelengths and colors. As a result, each element emits a unique color spectrum.

Bohr borrowed elements of Einstein's idea of light quanta and merged them with atomic spectra to propose a quantum mechanical model of atomic structure. The addition of exact energy levels provided the stability that the Rutherford model lacked. The next section details the absorption and emission of light according to the Bohr model of the atom.

Energy-Level Transitions

Niels Bohr's research involved the simplest atom possible, the hydrogen atom. This atom consists of a single proton and a single electron. A partial energy-level diagram of the hydrogen atom is shown in Figure 20.3.

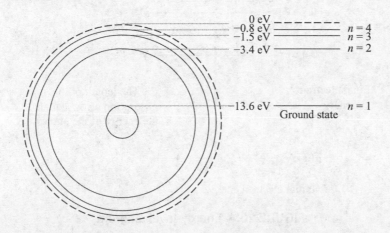

FIGURE 20.3 Energy levels of an atom

Although the actual model of the atom is not as simple as that shown in Figure 20.3, the Bohr model is still used in energy-level problems. The left side of the sketch shows a simplified Bohr model of a hydrogen atom. It shows the energy levels as circles similar to the orbits of planets in the solar system. The right side of the diagram shows the corresponding **energy-level diagram** of the atom. In an energy-level diagram, horizontal lines are used to portray the energy levels. The **ground state** is the lowest level an electron can occupy. It is numbered as the first energy level ($n = 1$). All the higher energy states are known collectively as the excited states. They are numbered from the ground state to the edge of the atom, $n = 2$, $n = 3$, etc. Be careful with the excited states. Since the ground state is $n = 1$, the first excited state is $n = 2$ and the second excited state is $n = 3$. The energy levels typically have negative values and are often measured in electron volts. The edge of the atom has a value of zero electron volts. The ground state for hydrogen has an energy of -13.6 electron volts. Moving deeper into the atom is similar to taking an elevator ride below ground. The floor numbers become larger the farther down the elevator travels, and the elevator is moving in the negative direction.

Absorption

Absorption occurs when a photon of light with the correct amount of energy enters an atom and is absorbed by an electron. The energy of the photon adds to the energy of the electron, creating an excited electron called a **photoelectron**. The high-energy photoelectron must move to a higher energy level. If the hydrogen atom is radiated by light with 12.1 electron volts of energy, the electron will absorb the photon of light and their energies will add.

$$E_{\text{electron initial}} + E_{\text{photon}} = E_{\text{electron final}} = (-13.6 \text{ eV}) + (12.1 \text{ eV}) = -1.5 \text{ eV}$$

Absorptions are indicated with upwardly drawn arrows in energy-level diagrams. The absorption calculated above is depicted in Figure 20.4. The photoelectron formed in this absorption moves upward 12.1 electron volts from the ground state to the third energy level, $n = 3$ (the second excited state).

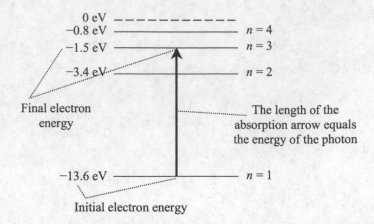

FIGURE 20.4 Energy-level diagram

To be absorbed, a photon must have an energy corresponding to the exact difference in energy levels in an atom.

Emission

Nature prefers low-energy states. So, electrons in high-energy levels will spontaneously move to lower energy levels until they finally reach the ground state. When electrons drop to a lower energy level, they lose energy. The energy they lose is given off as a photon of light. An **emission** is the light given off by an atom when electrons drop to lower energy levels. If an electron in the hydrogen atom at energy level two ($n = 2$) drops to the ground state ($n = 1$), it will lose 10.2 electron volts of energy.

$$E_{\text{photon}} = E_2 - E_1 = (-3.4 \text{ eV}) - (-13.6 \text{ eV}) = 10.2 \text{ eV}$$

The transitions to lower energy levels are haphazard. An electron in the third energy level might return all the way to the ground state in a single step. In a different atom, another electron in the third energy level might first drop to the second energy level and then drop to the ground state. Energy-level problems involve samples that contain vast quantities of atoms. All the possible drops between the energy levels take place in many different atoms simultaneously. Emissions of light are pictured as downward arrows in energy-level diagrams. Figure 20.5(a) shows all the possible emissions due to energy-level drops from energy level three. Figure 20.5(b) shows the possible emissions due to energy drops from energy level four.

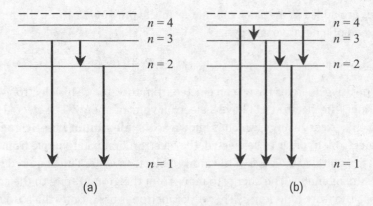

FIGURE 20.5 Emission of light

Ionization Energy/Work Function

In energy-level problems, electrons receive just enough energy to reach a higher energy level inside the atom. However, if the energy of the incoming photons is greater than the energy difference between the ground state and the edge of the atom, then the electrons are ejected from the atom. In this process, the atom becomes a positive ion. The minimum energy required to accomplish this is known as the **ionization energy**. For atoms with electrons in the ground state, the ionization energy is equal to the absolute value of the ground-state energy. For hydrogen gas with a ground state of -13.6 electron volts, the ionization energy is equal to 13.6 electron volts. The ionization energy is the minimum energy needed to eject an electron and ionize an atom. The ionization energy is also known as the **work function**, ϕ.

Photoelectric Effect

The photoelectric effect involves the ionization of atoms by striking them with photons that exceed the energy difference between the ground state and the edge of the atom. Figure 20.6 shows a hypothetical atom with a ground state of -10 electron volts that is radiated by photons with 12 electron volts of energy.

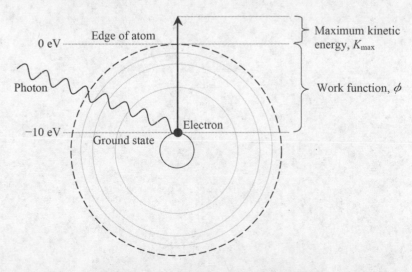

FIGURE 20.6 The photoelectric effect

If these energies are added in the same manner as in the previous section, the result is a positive energy instead of a negative energy.

$$E_{\text{electron initial}} + E_{\text{photon}} = E_{\text{electron final}} = (-10 \text{ eV}) + (12 \text{ eV}) = 2 \text{ eV}$$

Electrons with positive energies have been ejected from the atom. This electron will leave the atom. When it does, the electron will have 2 electron volts of energy. The ejected electron will be moving with this excess energy, which is known as the **maximum kinetic energy**, K_{max}, of the ejected electron. In order to be ejected, the electron first had to move from the ground state, -10 electron volts, to the edge of the atom, 0 electron volts. This required the addition of 10 electron volts of energy. The energy to move from the ground state to the edge of the atom is known as the work function, ϕ. The work function is essentially the absolute value of the ground-state energy. For the atom in Figure 20.6, the work function is

$$\phi = |E_{\text{ground state}}| = |-10 \text{ eV}| = 10 \text{ eV}$$

The excess energy of the ejected electrons can be determined as follows:

$$K_{\text{max}} = E_{\text{photon}} - \phi$$

The energy of a photon is related to its frequency, $E_{\text{photon}} = hf$. This expression can be substituted into the previous equation to complete the equation for the photoelectric effect.

$$K_{\text{max}} = hf - \phi$$

Both versions of this formula may be encountered. In Figure 20.6, the energy of the incident photon was given. So, the first formula, $K_{\text{max}} = E_{\text{photon}} - \phi$, is used.

$$K_{\text{max}} = (12 \text{ eV}) - (10 \text{ eV}) = 2 \text{ eV}$$

What is the significance of moving ejected electrons? When certain metallic substances with low work functions are radiated with high-energy photons, countless electrons are ejected. These electrons are in motion. A lot of moving electrons make up a current. As a result, this phenomenon is a way to generate an electric current using photons of light. It is known as the **photoelectric effect** since it converts photon energy to electric energy. This is how electricity is generated using sunlight. The photoelectric effect is at the heart of solar power.

CHAPTER 21

Nuclear Reactions

Learning Objectives

In this chapter, you will learn how to:

o Understand quantities used to understand nuclear reactions

o Identify subatomic particles involved in nuclear reactions

o Define isotopes and examine their importance in nuclear reactions

o Examine the nature of the forces acting on nuclear particles

o Understand the process of radioactive decay

o Differentiate between the key nuclear reactions of fission and fusion

Nuclear reactions involve particles in the nucleus of an atom. Nuclear reactions encompass a variety of reactions that lead to changes in the composition of the nucleus. Altering the number of protons in the nucleus of an atom causes the atom to become an entirely different element, a process known as **transmutation**.

Nucleons

Nuclear reactions are concerned with the nucleus of the atom and the particles it contains. The subatomic particles contained in the nucleus of an atom are known as **nucleons**. They consist of protons and neutrons. Both the mass and the number of these particles are important when analyzing nuclear reactions.

Atomic Mass Units

A specialized unit of mass known as the atomic mass unit (u) was devised to make working with the mass of fundamental particles easier. The mass of both the proton and neutron were originally thought to be the same. They were each assigned a mass of 1.0 atomic mass unit for simplicity. It has since been determined that the masses are very similar but that the neutron has a slightly greater mass. The modern definition of an atomic mass unit is $\frac{1}{12}$ the mass of a carbon-12 atom. By using this scale, a proton has a mass of 1.00728 atomic mass units, while a neutron has a mass of 1.00866 atomic mass units. For rough calculations, these masses are still both rounded off to 1.0 atomic mass unit. Atomic mass units are a more convenient scale to measure mass when working with fundamental particles.

Atomic Number and Mass Number

When the symbol for an element is used in a nuclear reaction, the atomic number and the mass number are written as subscripts and superscripts preceding the element's symbol, as shown in Figure 21.1.

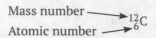

Mass number $\longrightarrow {}^{12}_{6}C$
Atomic number $\longrightarrow$

FIGURE 21.1 Atomic number and mass number

The **atomic number** is the number of protons in an atom. The number of protons defines an element. For example, all carbon atoms have 6 protons, so the atomic number of all carbon atoms is always 6. If a carbon atom gains or loses protons, it is no longer a carbon atom. Changing the number of protons causes a transmutation of the atom into a completely different type of element.

The **mass number** provides three different and important numerical values.

1. The mass number is the *number of protons plus neutrons*. In Figure 21.1, the mass number is 12. So, there are 12 protons and neutrons in a carbon-12 atom. Since the atomic number is the number of protons only, the number of neutrons can be deduced by subtracting the atomic number from the mass number. The carbon atom in Figure 21.1 has 6 neutrons ($12 - 6 = 6$).

2. The mass number is also the *atomic mass* (mass of a single atom) measured in atomic mass units. The carbon atom in Figure 21.1 has a mass of 12 atomic mass units.

3. In addition, the mass number is also the *molar mass* (mass of one mole of atoms) measured in grams per mole.

Subatomic Particles

Table 21.1 lists the fundamental particles most likely to be encountered in nuclear reactions. Knowing the mass and atomic numbers of these basic particles will be an asset when analyzing nuclear reactions. In addition, the charged particles will interact with electric and magnetic fields. You must know which particles are charged and whether that charge is positive or negative. Occasionally, problems ask how these particles move in electric and magnetic fields. Although you do not need to know the exact mass of the particles listed in the table, you should be able to list them in order of their masses.

TABLE 21.1 Subatomic Particles

Name	Symbol	Charge	Rest Mass (u)
Proton	$_1^1p$	+1	1.00728
Electron	$_{-1}^0e$	−1	0.00055
Neutron	$_0^1n$	0	1.00866
Neutrino	v_e	0	Nearly zero
Antineutrino	$\bar{v}_e$	0	Nearly zero
Alpha particle (helium nucleus)	$_2^4\alpha$ or $_2^4He$	+2	4.00150
Beta particle (an electron emitted from the nucleus)	$_{-1}^0\beta$ or $_{-1}^0e$	−1	0.00055
Gamma radiation	$_0^0\gamma$	0	0

In addition to protons, neutrons, and electrons, three new and important subatomic particles plus gamma radiation have been listed in the table.

Neutrino

Neutrinos are often a product of radioactive decay and nuclear reactions. A neutrino, v_e, is a neutral particle with very little (nearly zero) mass. Neutrinos are more common in the universe than electrons and protons. However, neutrinos do not interact well with matter, and they have insignificant mass. Neutrinos also have an antiparticle variant known as the antineutrino, $\bar{v}_e$. They have been included here as their symbols may be encountered in nuclear-reaction formulas.

Alpha Particle

An **alpha particle** is simply the nucleus of a helium atom without any electrons. The alpha particle has a mass number of 4. This means it contains 4 nucleons (protons + neutrons), and it has a mass of 4.0 atomic mass units. Of the particles listed in Table 21.1, the alpha particle is the most massive. The atomic number is 2, indicating that the alpha particle contains 2 protons. As a result, the alpha particle must also contain 2 neutrons. Since the alpha particle contains two protons and no electrons, it is positively charged. The charge of an alpha particle is equal to the charge of two protons ($+3.2 \times 10^{-19}$ coulombs). As a result, it interacts with electric and magnetic fields as would any positive charge.

Beta Particle

A **beta particle** is an electron produced when a neutron undergoes a transmutation to become a proton. In the process, the positive and negative charges cancel and the neutron becomes slightly more massive than the proton. Under the right conditions, a neutron may spontaneously divide and become a proton and an electron (plus an antineutrino, $\bar{v}_e$, which can be ignored).

$$_0^1n \rightarrow {}_1^1p + {}_{-1}^0e + \bar{v}_e$$

When this happens, the electron originates in the nucleus and not in the energy levels surrounding the nucleus, where ordinary electrons are found. This electron, known as a beta particle, is ejected from the atom with high energy.

Electrons and beta particles have too little mass to affect the atomic mass of an atom noticeably. Think of electrons as adding as much mass to an atom as eyelashes add to the mass of a person. The mass of electrons is therefore not included in the atomic mass.

Gamma Ray (Gamma Radiation)

Just like electrons, protons and neutrons can move between energy levels within the nucleus. Nuclear reactions can excite nucleons to higher energy levels. When the nucleons subsequently drop to lower energy levels, they emit photons. The photons emitted when nucleons drop to lower energy levels are incredibly energetic compared to those emitted when orbiting electrons change energy levels. These energetic photons are known as **gamma rays**, γ. Gamma rays are a form of electromagnetic radiation. They do not have mass or charge, and they are not influenced by electric or magnetic fields.

Isotopes

Although an atom must have a specific number of protons (a set atomic number) to remain a specific element, an atom does not need to contain a definite number of neutrons. Each element may have several combinations of neutrons that allow atoms of that element to exist in slightly different variations. As an example, carbon can be found in nature as $^{12}_{6}C$, $^{13}_{6}C$, and $^{14}_{6}C$, with 6, 7, and 8 neutrons, respectively. This means that although the atomic number of carbon is set, the mass number may vary. The forms of carbon with different numbers of neutrons are the possible **isotopes** of the carbon atom. Isotopes are the same chemical element but have different numbers of neutrons. The isotopes of an element are often reported with the element name followed by the mass number (carbon-12, carbon-13, and carbon-14). There is no need to report the atomic number (6) since it is a known fact that all carbon atoms have six protons.

The Strong Force

Students rarely ask why the protons in the nucleus cluster together when they should repel each other due to electrostatic forces. As it turns out, another force is operating in the nucleus, the **strong force**. The strong force attracts nucleons to one another. It attracts protons to protons, neutrons to neutrons, and protons to neutrons. While the electrostatic force is trying to separate the protons, the strong force holds them together. The magnitude of the strong force is greater than that of the electrostatic force. However, the strong force operates only at very small distances, such as those inside the nucleus.

Adding neutrons to the nucleus helps hold the nucleus together. The neutrons add to the strong force, helping to hold the repelling protons near each other. In addition, the neutrons have no charge, so they do not contribute to the electrostatic repulsive force that acts to tear apart the nucleus. However, every atom has an optimum mix of protons and neutrons. When the number of protons and neutrons falls outside of an optimum range, the geometry of the nucleus weakens the strong force. This allows the electrostatic force either to eject a small portion of the nucleus or to tear apart the entire nucleus. Carbon-14 and uranium-235 are classic examples of unstable nuclei. When these atoms undergo nuclear reactions, their nuclei experience changes in compositions. As a result, these elements undergo a transmutation into entirely new elements.

Mass-Energy Equivalence

During a nuclear reaction, the mass of the reactants at the start of the reaction does not equal the mass of the products produced by the reaction. The difference in mass between the products and the reactants is known as the **mass defect**, Δm. The mass defect is associated with the energy involved in a nuclear reaction. The amount of energy, E, associated with the mass defect can be calculated using the speed of light in a vacuum, c, and Albert Einstein's famous equation.

$$E = (\Delta m)c^2$$

This equation demonstrates **mass-energy equivalence**. During nuclear reactions, matter may be converted into energy or energy may be converted into matter.

Nuclear reactions release energy by converting a small amount of the original mass into energy. In nuclear reactions where energy is released, the reactants (ingredients) have more mass than the products formed during the reaction. To balance and account for all the original mass during the reaction, the mass defect must be added to the product side of the reaction.

$$\text{Reactants} \rightarrow \text{Products} + \Delta m$$

For example, a neutron splitting to form a proton and an electron can be shown as follows.

$$_0^1\text{n} \rightarrow {}_1^1\text{p} + {}_{-1}^{0}\text{e} + \overline{\nu}_e + \Delta m$$

$$1.00866 \text{ u} = 1.00728 \text{ u} + 0.00055 \text{ u} + \Delta m$$

$$\Delta m = 0.00083 \text{ u}$$

When the symbol for each particle is replaced by the appropriate mass in atomic mass units, it becomes apparent that the product side of the reaction is missing mass. The symbol for the mass defect is added to the side with the least amount of mass. So, the magnitude of the mass defect can be calculated.

The mass defect, Δm, is the small amount of mass that is converted into energy, $E = (\Delta m)c^2$, during the reaction. The values are shown here to assist you in understanding the concept of converting matter into energy.

$$\Delta m = 0.00083 \text{ u} \left(\frac{1.66 \times 10^{-27} \text{ kg}}{1 \text{ u}} \right) = 1.38 \times 10^{-30} \text{ kg}$$

$$E = (\Delta m)c^2 = (1.38 \times 10^{-30} \text{ kg})(3 \times 10^8 \text{ m/s})^2 = 1.24 \times 10^{-13} \text{ J}$$

This amount of energy is released when a neutron becomes a proton and an electron. Energy is a product of this reaction, and the mass defect can be replaced with energy in the reaction equation.

$$\ _0^1 n \rightarrow\ _1^1 p +\ _{-1}^0 e + \bar{\nu}_e + \text{Energy}$$

Radioactive Decay

Radioactive decay occurs when an unstable isotope spontaneously loses energy by emitting particles from its nucleus. The decay processes were named in their order of discovery by using the first three letters of the Greek alphabet: alpha (α), beta (β), and gamma (γ). After their initial discovery, it was determined that an alpha particle was in fact the nucleus of a helium atom ($_2^4$He), a beta particle was actually an electron ($_{-1}^0$e), and gamma radiation was not a particle at all. Instead, it is a high-frequency photon ($_0^0\gamma$). Elements that naturally decay are said to be radioactive. Radioactive substances have a critical imbalance between the number of protons and neutrons in the nucleus. As atoms become larger, more neutrons are needed to maintain stability. Uranium-235 has 92 protons and 143 neutrons. The three common forms of natural radioactivity—alpha, beta, and gamma radiation—are discussed below.

Alpha Decay

In an alpha decay, an alpha particle, $_2^4$He, is spontaneously ejected from the nucleus of an atom. If an alpha particle leaves the nucleus, the mass number of the atom is reduced by 4 while the atomic number is reduced by 2. Changing the atomic number causes a transmutation into a new element. The ejected alpha particle is the least dangerous form of radioactive decay. Although it is harmful if digested, an alpha particle can be stopped by both paper and skin.

EXAMPLE 21.1

The Product of an Alpha Decay

An isotope of uranium, $_{92}^{238}$U, undergoes alpha decay. In the process, the atom becomes an isotope of thorium. Which of the following elements is the result of this transmutation?

(A) $_{90}^{234}$Th   (B) $_{92}^{237}$Th   (C) $_{92}^{239}$Th   (D) $_{93}^{238}$Th   (E) $_{94}^{242}$Th

WHAT'S THE TRICK?

The particle resulting from the decay leaves the nucleus and is subtracted. Subtract the mass number (4) and atomic number (2) of the alpha particle from the original nucleus.

$$_{92}^{238}U \rightarrow\ _{92-2}^{238-4}Th +\ _2^4\alpha$$

$$_{92}^{238}U \rightarrow\ _{90}^{234}Th +\ _2^4\alpha$$

The resulting thorium nucleus has a mass number of 234 and an atomic number of 90. The answer is A.

Beta Decay

A beta particle is released when a neutron in the nucleus of an atom decays into a proton and an electron. This electron originates in the nucleus and is known as a beta particle. Beta particles move at greater speeds than alpha particles and can be stopped with a thin sheet of metal, such as aluminum.

EXAMPLE 21.2

The Product of Beta Decay

An isotope of carbon, $^{14}_{6}C$, undergoes beta decay. In the process, the atom becomes an isotope of nitrogen. Which of the following is the result of this transmutation?

(A) $^{10}_{4}N$ (B) $^{13}_{6}N$ (C) $^{15}_{6}N$ (D) $^{14}_{7}N$ (E) $^{18}_{8}N$

WHAT'S THE TRICK?

The particle resulting from the decay leaves the nucleus and is subtracted. Subtract the mass number (0) and atomic number (−1) of the beta particle from the original nucleus.

$$^{14}_{6}C \rightarrow \,^{14-0}_{6-(-1)}N + \,^{0}_{-1}\beta$$

$$^{14}_{6}C \rightarrow \,^{14}_{7}N + \,^{0}_{-1}\beta$$

The resulting nitrogen nucleus has a mass number of 14 and an atomic number of 7. The answer is D.

Gamma Rays

Gamma rays are the result of a number of radioactive decay reactions, including alpha and beta decay. Gamma rays travel at the speed of light and have greater penetration than alpha or beta radiation. Very dense materials, such as lead, are needed to stop gamma rays. The release of gamma rays (high-energy photons) does not affect the atomic number or mass number of the atoms.

Decay Rate

The rate at which radioactive decay occurs is often measured in what is known as a **half-life**. One half-life is the time interval needed for half of a sample of radioactive atoms to decay. Carbon-14, for example, has a half-life of 5,740 years (1 half-life = 5,740 years). At the end of this time period, only half of the original sample of carbon-14 remains. The rest has undergone a transmutation into nitrogen-14. Table 21.2 shows how a sample consisting of 16 grams of carbon-14 would progress through several half-lives.

TABLE 21.2 Decay of Carbon-14

Half-life	Time in Years	Carbon-14	Nitrogen-14
0	0	16 g	0 g
1	1(5,740) = 5,740 years	(½)(16) = 8 g	8 g
2	2(5,740) = 11,480 years	(½)(8) = 4 g	12 g
3	3(5,740) = 17,220 years	(½)(4) = 2 g	14 g
4	4(5,740) = 22,960 years	(½)(2) = 1 g	15 g

EXAMPLE 21.3

Half-Life

A 120-gram sample of iodine-131 has a half-life of 8.0 days. How much of the original sample remains after 24 days?

WHAT'S THE TRICK?

Divide the time interval by the length of time of one half-life to determine how many half-lives have passed.

$$(24 \text{ days})\left(\frac{1 \text{ half-life}}{8.0 \text{ days}}\right) = 3 \text{ half-lives}$$

During each half-life, the sample of iodine is halved. If 3 half-lives have passed, then the original 120-gram sample will be reduced by half 3 times.

$$(120 \text{ g})\left(\frac{1}{2}\right)\left(\frac{1}{2}\right)\left(\frac{1}{2}\right) = 120 \text{ g} \left(\frac{1}{2}\right)^3 = 15 \text{ g}$$

Fission and Fusion

A **fission** reaction is the splitting of a large atom into smaller atoms. A **fusion** reaction involves combining smaller atoms to make a larger atom. Spontaneous fission and fusion reactions involve the release of energy.

Fission

Fission of larger atoms into smaller ones is typically induced by the bombardment of the larger atom with free neutrons. The addition of a free neutron temporarily creates a larger, unstable nucleus. The attractive strong force is no longer able to hold the protons together. The repulsive electrostatic force tears apart the nucleus, forming two smaller nuclei. This process also releases several additional free neutrons and energy in the form of gamma rays. A common fission reaction occurring in nuclear power plants involves the splitting of the uranium-235 atom as shown in the following reaction.

$$^{235}_{92}\text{U} + ^{1}_{0}\text{n} \rightarrow ^{92}_{36}\text{Kr} + ^{141}_{56}\text{Ba} + 3(^{1}_{0}\text{n}) + \text{Energy}$$

In the example, the uranium-235 atom undergoes a transmutation into two distinct atoms, krypton and barium. It also releases three more free neutrons as well as energy. The energy released is used to heat water until it becomes steam. The steam is then used to rotate a coil of wire in a magnetic field, generating electrical energy. The three free neutrons are able to bombard three more uranium-235 atoms, causing additional fission reactions, which release even more neutrons. In power plants, the reaction is controlled. However, the number of free neutrons and the subsequent fission reactions have the potential to grow exponentially in what is known as a **chain reaction**.

The main difference between radioactive decay and fission is that fission requires activation and produces free neutrons to continue the reaction. Radioactive decay occurs spontaneously and produces no free neutrons. Both reactions release energy.

EXAMPLE 21.4

Fission

How many neutrons are created in the following nuclear reaction?

$$^{235}_{92}\text{U} + ^{1}_{0}\text{n} \rightarrow ^{90}_{38}\text{Sr} + ^{143}_{54}\text{Xe} + ?\,\text{n} + \text{Energy}$$

WHAT'S THE TRICK?

The mass numbers and atomic numbers must remain constant. The arrow separating the reactants and products can be treated as an equal sign.

$$\text{Mass numbers: } 235 + 1 = 90 + 143 + ?$$

$$\text{Atomic numbers: } 92 + 0 = 38 + 54 + ?$$

The mass numbers on the product side of the reaction are missing three atomic mass units. Each neutron has a mass number of one atomic mass unit, so this fission reaction must produce three free neutrons.

$$^{235}_{92}\text{U} + ^{1}_{0}\text{n} \rightarrow ^{90}_{38}\text{Sr} + ^{143}_{54}\text{Xe} + 3(^{1}_{0}\text{n}) + \text{Energy}$$

Fusion

The fusion of two smaller atoms to become a larger atom requires a tremendous amount of activation energy to overcome the electrostatic repulsion of the protons. Unlike larger atoms, such as uranium-235 or uranium-238, which have a number of protons close together, making it relatively easy to induce them to break apart, bringing small atoms together to make larger ones is quite difficult. An example of a fusion reaction is the fusion of two isotopes of hydrogen atoms into a helium atom.

$$^{2}_{1}\text{H} + ^{2}_{1}\text{H} \rightarrow ^{4}_{2}\text{He} + \text{Energy}$$

The isotopes of hydrogen in the above equation are known as **deuterium** and consist of 1 proton and 1 neutron. A fusion reaction will release more energy, per the mass of the reactants, than a fission reaction. However, inducing fusion is quite difficult because of the tremendous amount of energy needed to overcome the electrostatic repulsion of the two smaller reactant nuclei.

CHAPTER 22

Relativity

Learning Objectives

In this chapter, you will learn how to:

o Define the concept of special relativity and the laws of physics within a reference frame

o Determine what will happen to time, length, and mass as objects approach the speed of light

In 1905, Albert Einstein wrote four scientific papers that would change physics forever. One of those papers, on the photoelectric effect, would win him the Nobel Prize in 1921. Another paper, on special relativity, used creative-thought experiments to understand the nature and behavior of the speed of light and objects traveling near such speeds. Although his ideas are complex and often seem counterintuitive, observations of testable phenomena have consistently agreed with his theory of special relativity.

Special Theory of Relativity

Until Einstein's paper in 1905, most physicists believed that the universe was filled with an invisible medium called ether. They reasoned that ether was necessary in order for light, considered by most physicists to be a wave, to propagate through space. However, experiments such as the famous Michelson-Morley experiment all failed to prove the existence of ether.

Einstein viewed the problem in a completely different manner and proposed an explanation for these failed experiments. He viewed light as a quantum particle (later named a photon) that traveled with a specific speed in a vacuum, $c = 3 \times 10^8$ meters per second. Einstein then suggested that the speed of light is the same for all **inertial reference frames**. An inertial reference frame is a frame of reference moving at a constant velocity. (Inertia is the tendency of objects to continue moving at constant velocity.) Einstein stated that all the laws of physics are the same for any inertial reference frame. This is known as Einstein's **first postulate of special relativity**.

In addition to the speed of light's being the same for all inertial reference frames, Einstein stated that even if light were emitted from a moving source, it would continue to have a velocity of c. This is known as Einstein's **second postulate of special relativity**. The second postulate is counterintuitive to the laws of motion that describe the relative motion of two moving objects. Consider a ball being thrown forward from a moving vehicle. The true speed of the ball is the speed of the throw plus the speed of the moving vehicle. This is not so with light. Light beams from a car's headlight travel at c regardless of the speed of the car. The true speed of light is always c when measured in any inertial reference frame.

Einstein's special relativity is a special case since inertial reference frames are not accelerating. In 1915, Einstein proposed a theory of general relativity that encompassed accelerating reference frames in addition to inertial reference frames.

Time, Length, and Mass

Using the assumption that light travels at a constant velocity of c for any inertial reference frame, several noticeable effects will happen to time, length, and mass at speeds approaching c. Note that the formulas in the following sections are provided for context.

Time Dilation

Picture a person watching a moving object. An observer will usually interpret his or her own motion as being stationary and will perceive the other object as moving. Both the observer (stationary) and the object (moving at constant velocity) are equipped with clocks. The observer will see his or her own clock as running normally but will see the clock on the moving object as running slowly. This effect is known as **time dilation**. The amount that the clock appears to be running slowly can be calculated with the following equation.

$$t = \frac{t_0}{\sqrt{1 - v^2/c^2}}$$

Time, t_0, is the time on the clock of the stationary observer, which appears to be running normally. However, when the stationary observer looks at the clock on the moving object, it will have a different, slower time, t. The velocity of the moving object is v, and c is the speed of light. If the moving object has a small velocity, both clocks will appear to be running at the same time, $t \approx t_0$. The time difference between the two clocks will be negligible, and the effects of time dilation will go unnoticed.

$$t = \frac{t_0}{\sqrt{1 - v^2/c^2}} \approx \frac{t_0}{\sqrt{1 - 0}} \approx t_0$$

The speeds of manmade objects are too slow for dilation effects to appear. As a result, we are unaware that dilation actually occurs. However, as the speed of the object nears the speed of light, the effects of time dilation increase dramatically. When an object moves at 99.5 percent of the speed of light ($v = 0.995c$), the time difference in the clocks is substantial.

$$t = \frac{t_0}{\sqrt{1 - \dfrac{v^2}{c^2}}} \approx \frac{t_0}{\sqrt{1 - \dfrac{(0.995c)^2}{(1.000c)^2}}} \approx 10t_0$$

Length Contraction

When an object moves, the length of the object and anything moving with the object appears to contract in length. The observed length is calculated with the following equation.

$$L = L_0\sqrt{1 - v^2/c^2}$$

You should note that the actual length affected by motion is the length that matches the direction of motion. If an object is moving to the right, only length along the x-axis is affected. The height of the object in the y-direction and its depth in the z-direction remain unchanged. In the formula above, length L_0 is the object's rest length. This is the length of the object measured when the object is at rest or by someone moving at the same speed as the object. Length L is the length measured by a stationary observer, someone not moving with the object. At low velocities, v, these lengths are nearly identical, $L \approx L_0$, and the length contraction is not noticeable.

$$L = L_0\sqrt{1 - v^2/c^2} = L_0\sqrt{1 - 0} = L_0$$

Momentum Effects

Momentum, like time and length, is also affected by the motion of objects. At slow speeds the momentum of an object is directly proportional to its velocity, $p = m_0 v$, where m_0 is an object rest mass. The rest mass is the mass of the object when it is stationary. However, as the speed of an object approaches c, the momentum of the object increases in a dramatic nonlinear manner.

$$p = \frac{m_0 v}{\sqrt{1 - v^2/c^2}}$$

As with time and length, the effects are unnoticed at the low speeds that dominate human experience.

$$p = \frac{m_0 v}{\sqrt{1 - v^2/c^2}} = \frac{m_0 v}{\sqrt{1 - 0}} = m_0 v$$

Index

A

Absorption, 154, 176–178
Acceleration
  centripetal, 48–50
  constant, 18
  formula, 17
  of gravity, 18–19
  kinematic quantities, 17
  Newton's second law, 36–37
  oscillations, 123
  on velocity, 17–18
  velocity vector, 9–11
Action force, 37
Adding, vectors, 12–14
Adiabatic process, 170
Agent, in force, 31
Alpha decay, 186
Alpha particle, 183
Amplitude, 118, 125
Angle of reflection, 136
Angular velocity, 51
Antinodes, 132
Applied force, 32
Area, calculating, 4
Area and slope related to time,
  formula, 6
Atomic mass unit, 181
Atomic number, 182
Atomic theory
  electromagnetic radiation,
   176
  electron volts, 175
  energy level transitions,
   177–179
  gold-foil experiment,
   174–175
  ionization energy, 179
  light wavelengths, 176
  photoelectric effect, 179–180
  raisin cake model, 173–174
  spectroscopy, 175
  variables used in, 173
  work function, 179
Atwood, George, 44
Atwood machine, 44

B

Battery, 92–93, 95–96
Beats, 133–134
Beta decay, 187
Beta particle, 183–184
Bohr, Niels, 176

C

Capacitance, 85, 92
Capacitors, 92–94
Carbon-12, 181
Carbon-14, 187–188
Celsius temperature, 158
Centripetal acceleration, 48–50
Centripetal force, 50
Chain reaction, 189
Charge, 76–77, 90
Circuit elements
  DC circuit, 95–102
  heat and power dissipation,
   102–103
Circular motion, 46–50
Circular orbits, 72–74
Circular wave fronts, 149
Coefficient of kinetic friction,
  33
Coefficient of static friction, 33
Collisions, 67–68
Color, 154–155
Compound-body problems,
  41–44
Concave lens, 143–144
Concave mirrors, 144
Conduction, 77, 163
Conductor, 77
Conservation of energy, 62–63,
  91
Conservation of momentum,
  67–69
Constant acceleration, 18
Constant force, work done by,
  56–57
Constant speed formula, instan-
  taneous velocity vectors, 48
Constant velocity formula, 19

Constructive interference, 131,
  150
Convection, 163, 164
Convection current, 164
Converging lenses, 140–143
Converging mirrors, 144–146
Convex lens, 142–143
Convex mirrors, 144
Coordinate system, 7–8
Coulomb's law, 83
Current, 96
Current-carrying wires, 106,
  112–113
Cycle, in uniform circular mo-
  tion, 47–48

D

DC circuit
  battery, 95–96
  equivalent resistance, 97
  parallel circuits, 100–102
  resistors, 96
  series, 97–100
  switch, 97
Decay rate, 187–188
Derived units, 1–2
Destructive interference, 131,
  150
Deuterium, 189
Diffraction, 148–149
Diffuse reflection, 136
Direction, vectors, 8–14
Dispersion, of light, 155
Displacement
  angular, 51
  graph of force vs, 59–60
  kinematic equation, 21
  kinematic quantities, 16
  true velocity, 24–25
  vectors, 55–56
  velocity vector, 9–11
Distance, kinematic quantities,
  17
Diverging lenses, 143–144
Diverging mirrors, 144–147
Domains, 105

Doppler-effect, 128–130
Double-slit experiment, 150–152
Dynamic equilibrium, 36
Dynamics, kinematics and,
  40–41

E

Einstein, Albert, 176, 190
Elastic collisions, 68
Elastic potential energy, 53–54
Electric fields, 75–77, 80–83,
  105
Electric potential
  capacitors and, 92–94
  energy, 89–91
  motion of charges and, 90
  of point charges, 88
  of uniform fields, 86–87
  variables used in, 85
Electricity, 90–91, 116
Electromagnet, 108
Electromagnetic induction,
  113–116
Electromagnetic radiation, 176
Electromagnetic spectrum, 128
Electromagnetic waves,
  127–128
Electron volt, 175
Electrostatic force, 184
Electrostatic repulsion, 189
EMF, 113, 115–116
Emission, 176, 178–179
Energy
  of capacitors, 93
  in collisions, 67
  conservation of, 62–63, 91
  electric potential, 89–90
  internal, 167–168
  ionization, 179
  level transition, 177–179
  levels, 176–178
  mechanical, 52–54
  model, 170
  oscillations, 122–123
  potential, 53–54
  quantum of, 176

Energy (*continued*)
    transfer, 168–170
    variables used in, 52
Entropy, 171
Equilibrium, 36
Equilibrium position, 119
Equipotential lines, 86–87
Equivalent resistance, 97
External forces, 41

F

Faraday, Michael, 115
Filter, polarizing, 153
First harmonic waveform,
    132–133
First law of thermodynamics,
    170
First postulate of special rela-
    tivity, 190
Fission, 188–189
Fixed magnets, 105–106,
    112–113
Force
    action, 37
    applied, 32
    centripetal, 50
    common, 31
    conservative, 62
    diagrams, 35–36
    direction, 31
    displacement graph, 59–60
    external, 41
    formula, 2
    of friction, 33
    of gravitational attraction,
        70–71
    of gravity (weight), 32
    internal, 41
    on moving charges,
        108–111
    nonconservative, 62–63
    normal, 32–33
    oscillations, 123
    parallel, 55
    restoring force, 35
    solving problems, 38
    superposition of, 84
    velocity vector, 9–11
    work done by magnetic, 111
Frame of reference, coordinate
    system, 8
Free-body diagram, 33, 35
Frequency, 47–48, 117, 126, 134
Friction, 33–34, 62, 77, 157–158

Function, slope of a graphed, 3
Fundamental frequency,
    132–133
Fundamental metric units (SI
    units), 1
Fusion, 188–189

G

Gamma ray (radiation), 184, 187
Geometric optics, variables used
    in, 135
Gold-foil experiment, 174–175
Graphs
    force vs displacement,
        59–60
    force-time, 66–67
    interpreting, 5–6
    kinematic, 21–22
    simple harmonic motion,
        122
    variables, 2–3
Gravitational attraction, force
    of, 70–71
Gravitational field, 71–72
Gravitational potential energy,
    53, 62
Gravity
    gravitational field, 71–72
    inverse-square law, 71
    surface, 72
    universal, 70–71
    variables used in, 70
    work by, 56–57
Ground state, 177

H

Half-life, 187–188
Harmonics, 132–133
Heat, 102, 162, 169–170
Heat of fusion, 165
Heat of vaporization, 165–166
Heat reservoir, 169
Heat sink, 165
Heat transfer, 162–164
Heat transformation, 165
Heating and cooling, 164–166
Hooke's Law
    elastic potential energy, 53
    in simple harmonic motion,
        119
    in springs, 35
    work by springs, 57–59
Horizontal displacement, vec-
    tors in, 10

Hot, 162
Huygens, Christian, 149
Huygens' principle, 149
Hydrogen atom, 177

I

Ideal gas law, 160–162
Imaginary positive test charge,
    81
Impulse, 65–67
Impulse-momentum theorem,
    65–66
Incident ray, 136
Inclines, 38–40
Index of refraction, 137–139
Induction, 77
Inelastic collisions, 68
Inertia, 31, 36
Inertial reference frames, 190
Instantaneous velocity vectors,
    48
Insulator, 77
Interference of light, 150
Internal energy, 167–168
Internal forces, 41
Inverse tangent, component
    vectors, 12
Inverse-square law, 71, 81
Ion, 76
Ionization energy, 179
Isobaric, 169
Isolated system, 172

J

Joule's law, 102

K

Kelvin temperature, 158
Kepler, Johannes, 74
Kepler's laws, 74
Kinematics
    dynamics and, 40–41
    equations, 19–21, 24–25
    graphs, 21–22
    one dimension, 16–22
    two dimension, 23–29
    uniform electric field,
        79–80
Kinetic energy
    lost, 60–63
    mechanical, 53
    work-kinetic energy
        theorem, 60–61
Kinetic-friction, 34

L

Latent heat, 165
Law of inertia, 31, 36
Law of reflection, 136–137
Laws of motion, Atwood ma-
    chine, 44–45
Length contraction, 192
Lens
    converging convex,
        140–143
    diverging concave, 143–144
Lenz's law, 116
Light
    absorption, 154, 176–178
    bulbs, 96
    double-slit experiment,
        150–152
    interference of, 150
    mediums, 135
    polarization of, 152–154
    ray model of, 136
    reflection, 136
    time, length, and mass,
        191–192
    wavelengths, 176
    waves, 126–128
Linear expansion, 159
Linear momentum, 64
Linear-motion problems, 50
Longitudinal wave, 124–125

M

Magnetic flux, 113–115
Magnetism
    current-carrying wires,
        106–107
    Electromagnet, 108
    field vs electric field, 105
    fixed, 105–106
    force on current-carrying
        wires, 112–113
    force on moving charges,
        108–111
    Lenz's law, 116
    magnitude of field of wire,
        107
    moving charges, 108–111
    permanent, 105–106
    solenoid, 108
    uniform magnetic fields,
        108–111
    variables used in, 104
    work by magnetic force, 111

Magnitude
  of electric field of point
    charges, 81–82
  EMF, 115–116
  of field of wire for magnets,
    107
  scalar, 8
  solving for, 41–42
  vector, 8–11
Mass defect, 185
Mass number, 182
Mass-energy equivalence, 185
Maximum kinetic energy, 180
Mechanical energy
  converting thermal energy
    to, 157–158
  forms of, 52–54
  kinetic energy in, 53
Mechanical waves, 127
Medium
  light, 128
  sound, 127
  waves, 126
Mirrors, 137, 144–147
Momentum, 64–65, 67–69, 192
Monochromatic light, 150
Motion
  circular, 46–50
  independence of, 23–24
  kinematic equations, 24
  laws of, 44–45
Motional emf, 116

N

Natural radioactivity, 186–188
Net force, 31
Neutrino, 183
Newton, Isaac, 36
Newton's first law, 36
Newton's second law, 36–37
Newton's third law, 32, 37, 43
Newton's universal law of gravi-
    tation, 70–72
Nodes, standing, 132
Normal line, 136
Nuclear reactions
  atomic mass unit, 181
  atomic number, 182
  decay rate, 187–188
  fission, 188–189
  fusion, 189
  isotopes, 184
  mass-energy equivalence,
    185–186

radioactive decay, 186–188
strong force, 184–185
subatomic particles,
  182–184
transmutation, 181

O

Objects, neutral, 76
Ohmic resistors, 96
One dimension
  force variables and units, 30
  kinematics, 16–22
  Newton's second law, 36–37
Orbital velocities, calculating,
  73–74
Ordinary collisions, 69
Oscillations
  acceleration, 123
  electromagnetic field, 128
  energy, 122–123
  pendulum, 121–122
  springs, 118–121
  waves, 124

P

Parallel circuits, 100–102
Parallel force, 55
Parallelogram method, adding
    vectors, 14–15
Path difference, 152
Pendulum, oscillations of,
  121–122
Perfectly inelastic collisions, 69
Period, 47–48, 117
Permanent magnets, 105–106
Phase changes, 165
Photoelectric effect, 179–180,
  190
Photoelectron, 177
Photons, 176
Physical optics
  diffraction, 148–149
  interference of light,
    150–152
  polarization of light,
    152–154
  variables used in, 148
Pitch, of wave, 126
Planck, Max, 175
Planck's constant, 175
Plane mirror, 137
Plane of polarization, 153
Plum-pudding model, 173–174
Point charges, 80–81, 88

Polarization of light, 152–154
Polarizing filter, 153
Potential difference, 90
Potential energy, 53–54
Power, 52, 61, 103
Power dissipation, 102
Pressure, ideal gas, 161–162
Projectile motion, 10, 27–29
Pulley problems, 44–45
Pythagorean theorem, compo-
    nent vectors, 12

Q

Quantum phenomena, variables
  used in, 173

R

Radiation, 163, 164
Radioactive decay, 183, 186–188
Raisin cake model, 173–174
Ray model of light, 136
Reaction force, 37
Reference frames, 190
Reflected ray, 136
Reflection, 136–137, 154
Refraction, 137–139
Relative-velocity equations, 25
Relativity, 190–192
Resistance, 96
Resistors, 96
Restorative force, 119
Restoring force, 35
Right-hand rule, 106, 110
Rutherford, Ernest, 174–175

S

Scalar, commonly used, 8
Scattering, of light, 155
Second law of thermodynamics,
  172
Second postulate of special
    relativity, 190
Second-harmonic waveform,
  133
Series circuits, solving for, 97–100
SI units, 1
Sign conventions, on kinematic
    quantities, 18–19
Simple harmonic motion
  force and acceleration, 123
  graphical representations
    of, 122
  oscillations of pendulum,
    121–122

oscillations of springs,
  118–121
terms, 117–118
Sinusoidal pattern, 125
Slope, of a graphed function, 3
Snell's law, 138–139
Solar power, 180
Solenoid, 108
Sonic boom, 130
Sound barrier, 130
Sound waves, 126, 127
Special theory of relativity,
  190–191
Specific heat, 164
Spectroscopy, 175
Specular reflection, 136
Spherical mirrors
  concave mirrors, 144
  converging, 144–146
  convex mirrors, 144
  overview of, 144
Spring constant, 35, 118–119
Springs
  determining the period,
    119–121
  force, 35
  Hooke's Law, 57–59
  oscillations of, 118–121
  work by, 57–59
Standing waves, 131–132
Static equilibrium, 36
Static friction, 33–34
Stationary object, Doppler-ef-
    fect, 130
Subatomic particles, 182–183
Subsonic object, 130
Superposition, 82–84, 131
Supersonic, 130
Surface gravity, 72
Switch, 97

T

Tangential velocity, 48–50,
  73–74
Temperature, 158
Tension, 34
Thermal conductivity,
  163–164
Thermal energy, 157–158
Thermal equilibrium, 163
Thermal expansion, 158–159
Thermal properties
  heat, 162
  heat transfer, 162–164

Thermal properties (*continued*)
  heating and cooling, 164–166
  ideal gas, 160–162
  linear expansion, 159
  mechanical energy, 157–158
  overview of, 156
  temperature, 158
  thermal energy, 157
  thermal expansion, 158–159
Thermodynamics
  energy model, 170
  energy transfer in, 168–170
  entropy, 171
  first law of, 170
  internal energy, 167–168
  second law of, 172
  variables used in, 167, 170
Thin lenses
  converging lenses, 140–143
  convex lens, 139–140
  diverging lenses, 143–144
  refraction, 139
Thomson, J.J., 173
Time, 19–21, 191
Tip-to-tail method, adding vectors, 12–14
Total mechanical energy, 54
Total momentum, 65
Transmutation, 181
Transverse wave, 124–125
Trigonometry, right-triangle, 11
True displacement, 25
True velocity, 25–26
Two dimensions
  displacement, 24–25
  kinematics, 23–29
  motion, 23–24
  velocity, 24–26

U

Uniform acceleration, 18
Uniform circular motion, period and frequency, 47–48
Uniform electric fields
  electric force in, 77–78
  kinematics in, 79–80
  magnitude of, 77
  potential of, 86–87
  vs uniform gravitational field, 77
Uniform gravitational field, vs uniform electric fields, 77
Uniform magnetic fields, 108–111
Units. *See* Variables
Universal gravity, 70–71
Universal law of gravitation, 70–72
Uranium-235, 186, 188–189

V

Variables, 135
  atomic theory, 173
  circuits, 95
  circular motion, 46
  dynamics, 30
  electric fields, 75
  electric potential, 85
  energy, work, and power, 52
  geometric optics, 135
  graphing, 2–3
  gravity, 70
  impulse, 64
  kinematic, 18–19
  magnetism, 104
  momentum, 64
  one dimension, 30
  physical optics, 148

  quantum phenomena, 173
  thermal energy, 157
  thermal properties, 157
  thermodynamics, 167, 170
  in two-dimensional kinematics, 23
  waves, 124
Vectors
  adding, 12–14
  commonly used, 9
  direction, 8–14
  inverse tangent, 12
  magnitude, 11
  mathematics, 11–15
  missing, 15
  overview of, 8–11
  parallelogram methods, 14–15
  tip-to-tail method, 12–14
Velocity
  acceleration effect, 17–18
  angular, 51
  constant, 19
  displacement and true, 24–25
  kinematic equation, 20
  kinematic quantities, 17
  Newton's second law, 36–37
  projectile, 27
  relative, 25–26
  tangential, 48–50
  time versus, 5
  true, 24–25
  vectors, 9–11, 48
Vertical compound-body problems, 42
Virtual image, 137
Visualizing
  field of current-carrying wires, 106–107

  uniform fields, 78, 80–81
Voltage, 98–99

W

Wavelength, 125
Waves
  amplitude effect on, 126
  components, 125
  electromagnetic, 127–128
  front model, 128–129
  fronts, 149
  mechanical, 127
  oscillations in, 124
  sound, 126
  speed, 126
  traveling, 124–125
  variables used in, 124
  See also Physical optics
Weight, of an object, 32
Work
  of electricity, 90–91
  energy transfer, 168–169
  force and displacement, 55–56
  by gravity, 56–57
  by magnetic force, 111
  by spring, 57–59
  variables used in, 52
Work function, 179
Work-kinetic energy theorem, 60–61

Y

Young, Thomas, 150
Young's double-slit experiment, 150–152

Z

Zero-reference potential, 86